COMPUTING RESEARCH FOR SUSTAINABILITY

Lynette I. Millett and Deborah L. Estrin, *Editors*

Committee on Computing Research for
Environmental and Societal Sustainability

Computer Science and Telecommunications Board

Division on Engineering and Physical Sciences

NATIONAL RESEARCH COUNCIL
OF THE NATIONAL ACADEMIES

THE NATIONAL ACADEMIES PRESS
Washington, D.C.
www.nap.edu

THE NATIONAL ACADEMIES PRESS   500 Fifth Street, NW   Washington, DC 20001

NOTICE: The project that is the subject of this report was approved by the Governing Board of the National Research Council, whose members are drawn from the councils of the National Academy of Sciences, the National Academy of Engineering, and the Institute of Medicine. The members of the committee responsible for the report were chosen for their special competences and with regard for appropriate balance.

Support for this project was provided by the National Science Foundation under award 115-0950451. Any opinions, findings, conclusions, or recommendations expressed in this publication are those of the authors and do not necessarily re ect the views of the organization that provided support for the project.

International Standard Book Number-13: 978-0-309-25758-9
International Standard Book Number-10: 0-309-25758-1

Copies of this report are available from:

The National Academies Press
500 Fifth Street, NW, Keck 360
Washington, DC 20001
(800) 624-6242 or
(202) 334-3313
http://www.nap.edu

Copyright 2012 by the National Academy of Sciences. All rights reserved.

Printed in the United States of America

THE NATIONAL ACADEMIES
Advisers to the Nation on Science, Engineering, and Medicine

The **National Academy of Sciences** is a private, nonprofit, self-perpetuating society of distinguished scholars engaged in scientific and engineering research, dedicated to the furtherance of science and technology and to their use for the general welfare. Upon the authority of the charter granted to it by the Congress in 1863, the Academy has a mandate that requires it to advise the federal government on scientific and technical matters. Dr. Ralph J. Cicerone is president of the National Academy of Sciences.

The **National Academy of Engineering** was established in 1964, under the charter of the National Academy of Sciences, as a parallel organization of outstanding engineers. It is autonomous in its administration and in the selection of its members, sharing with the National Academy of Sciences the responsibility for advising the federal government. The National Academy of Engineering also sponsors engineering programs aimed at meeting national needs, encourages education and research, and recognizes the superior achievements of engineers. Dr. Charles M. Vest is president of the National Academy of Engineering.

The **Institute of Medicine** was established in 1970 by the National Academy of Sciences to secure the services of eminent members of appropriate professions in the examination of policy matters pertaining to the health of the public. The Institute acts under the responsibility given to the National Academy of Sciences by its congressional charter to be an adviser to the federal government and, upon its own initiative, to identify issues of medical care, research, and education. Dr. Harvey V. Fineberg is president of the Institute of Medicine.

The **National Research Council** was organized by the National Academy of Sciences in 1916 to associate the broad community of science and technology with the Academy's purposes of furthering knowledge and advising the federal government. Functioning in accordance with general policies determined by the Academy, the Council has become the principal operating agency of both the National Academy of Sciences and the National Academy of Engineering in providing services to the government, the public, and the scientific and engineering communities. The Council is administered jointly by both Academies and the Institute of Medicine. Dr. Ralph J. Cicerone and Dr. Charles M. Vest are chair and vice chair, respectively, of the National Research Council.

www.national-academies.org

COMMITTEE ON COMPUTING RESEARCH FOR ENVIRONMENTAL AND SOCIETAL SUSTAINABILITY

DEBORAH L. ESTRIN, University of California, Los Angeles, *Chair*
ALAN BORNING, University of Washington
DAVID CULLER, University of California, Berkeley
THOMAS DIETTERICH, Oregon State University
DANIEL KAMMEN, University of California, Berkeley
JENNIFER MANKOFF, Carnegie Mellon University
ROGER D. PENG, Johns Hopkins Bloomberg School of Public Health
ANDREAS VOGEL, SAP Labs

Staff

LYNETTE I. MILLETT, Senior Program Officer
VIRGINIA BACON TALATI, Associate Program Officer
SHENAE BRADLEY, Senior Program Assistant

COMPUTER SCIENCE AND TELECOMMUNICATIONS BOARD

ROBERT F. SPROULL, Oracle (retired), *Chair*
PRITHVIRAJ BANERJEE, ABB
STEVEN M. BELLOVIN, Columbia University
JACK L. GOLDSMITH III, Harvard Law School
SEYMOUR E. GOODMAN, Georgia Institute of Technology
JON M. KLEINBERG, Cornell University
ROBERT KRAUT, Carnegie Mellon University
SUSAN LANDAU, Radcliffe Institute for Advanced Study
PETER LEE, Microsoft Corporation
DAVID LIDDLE, U.S. Venture Partners
DAVID E. SHAW, D.E. Shaw Research
ALFRED Z. SPECTOR, Google, Inc.
JOHN STANKOVIC, University of Virginia
JOHN SWAINSON, Silver Lake Partners
PETER SZOLOVITS, Massachusetts Institute of Technology
PETER J. WEINBERGER, Google, Inc.
ERNEST J. WILSON, University of Southern California
KATHERINE YELICK, University of California, Berkeley

Staff

JON EISENBERG, Director
RENEE HAWKINS, Financial and Administrative Manager
HERBERT S. LIN, Chief Scientist
LYNETTE I. MILLETT, Senior Program Officer
EMILY ANN MEYER, Program Officer
VIRGINIA BACON TALATI, Associate Program Officer
ENITA A. WILLIAMS, Associate Program Officer
SHENAE BRADLEY, Senior Program Assistant
ERIC WHITAKER, Senior Program Assistant

For more information on CSTB, see its web site at http://www.cstb.org, write to CSTB, National Research Council, 500 Fifth Street, NW, Washington, DC 20001, call (202) 334-2605, or e-mail the CSTB at cstb@nas.edu.

Preface

Computer science and information technologies offer a wide range of tools for examining sustainability challenges. Advances in computer science have already provided environmental and sustainability researchers with a valuable tool set—computational modeling, data management, sensor technology, machine learning, and other tools—and additional research in computer science may provide advanced approaches, tools, techniques, and strategies toward understanding, addressing, and communicating sustainability challenges.

The present study emerged from an informal request to the National Research Council's Computer Science and Telecommunications Board (CSTB) from the Directorate for Computer and Information Science and Engineering, National Science Foundation (NSF). The project was funded by the National Science Foundation. The statement of task for the Committee on Computing Research for Environmental and Societal Sustainability, established by the National Research Council to carry out this study, is as follows:

> Computing has many potential "green" applications including improving energy conservation, enhancing energy management, reducing carbon emissions in many sectors, improving environmental protection (including mitigation and adaptation to climate change), and increasing awareness of environmental challenges and responses. An ad hoc committee would plan and conduct a public workshop to survey sustainability challenges, current research initiatives, results from previously-held topical workshops, and related industry and government development

efforts in these areas. The workshop would feature invited presentations and discussions that explore research themes and specific research opportunities that could advance sustainability objectives and also result in advances in computer science and consider research modalities, with a focus on applicable computational techniques and long-term research that might be supported by the National Science Foundation, and with an emphasis on problem- or user-driven research.

The committee would obtain additional inputs through briefings to the committee and solicitations of comments and white papers from the research community. It would use additional deliberative meetings of the committee to develop a consensus report identifying promising research opportunities, cataloging applicable computational techniques, laying out an overall framework for "green" computing research, and recommending long-term research objectives and directions. The committee's consensus report will include a summary of the workshop as an appendix.

The committee reviewed current efforts underway in industry (and other opportunities for the immediate application of existing information technology) and explored research themes and specific research opportunities that could advance sustainability (energy and environmental) objectives and also result in advances in computer science. The committee considered research modalities, with a focus on applicable computational techniques and long-term research.

The report, which includes as Appendix A the summary of the Workshop on Innovation in Computing and Information Technology for Sustainability, identifies promising research opportunities, catalogs applicable computational techniques, lays out an overall framework for computing research for sustainability, and recommends long-term research objectives and directions. Chapter 1 provides examples of domains of potential impact, Chapter 2 describes methods and approaches, and Chapter 3, which is aimed primarily at computer science researchers, articulates why the interplay between addressing sustainability challenges and computer science research merits attention.

Meeting these challenges will involve advances in a number of computing research areas, including the following: scalability; robustness; reliability; real-time observation and processing; low-power computing, and sensing and actuation; and human interaction with the environment, observations, and feedback systems. A number of specific areas of computer science and topics addressed in current research programs of NSF's Directorate for Computer and Information Science and Engineering are relevant.

This report represents the cooperative effort of many people. The members of the study committee, after substantial discussions, drafted and worked through several revisions of the report. The committee would like to thank Jeannette Wing, Sampath Kannan, and Douglas Fisher for their encouragement and support of this study. The committee also appreciates the insights and perspective provided by the following experts who presented briefings:

> Adjo Amekudzi, Georgia Institute of Technology,
> Peter Bajcsy, National Institute of Standards and Technology,
> Eli Blevis, Indiana University, Bloomington,
> David Brown, Duke University,
> Randal Bryant, Carnegie Mellon University,
> David Douglas, National Ecological Observatory,
> John Doyle, California Institute of Technology,
> Chris Forest, Pennsylvania State University,
> Thomas Harmon, University of California, Merced,
> Neo Martinez, Pacific Ecoinformatics and Computational Ecology Lab,
> Vijay Modi, Columbia University,
> Shwetak Patel, University of Washington,
> Robert Pfahl, International Electronics Manufacturing Initiative,
> David Shmoys, Cornell University, and
> Bill Tomlinson, University of California, Irvine.

Finally, I thank CSTB staff members Lynette Millett and Virginia Bacon Talati for their efforts in steering the committee's work, coordinating the meetings and speakers, and drafting, editing, and revising report material.

Deborah L. Estrin, *Chair*
Committee on Computing Research for
Environmental and Societal Sustainability

Acknowledgment of Reviewers

This report has been reviewed in draft form by individuals chosen for their diverse perspectives and technical expertise, in accordance with procedures approved by the National Research Council's Report Review Committee. The purpose of this independent review is to provide candid and critical comments that will assist the institution in making its published report as sound as possible and to ensure that the report meets institutional standards for objectivity, evidence, and responsiveness to the study charge. The review comments and draft manuscript remain confidential to protect the integrity of the deliberative process. We wish to thank the following individuals for their review of this report:

Alice Agogino, University of California, Berkeley,
Ruzena Bajcsy, University of California, Berkeley,
Jeff Dozier, University of California, Santa Barbara,
Brian Gaucher, T.J. Watson Research Center, IBM,
Roger Ghanem, University of Southern California,
Marija Ilic, Carnegie Mellon University,
David Shmoys, Cornell University, and
Bill Tomlinson, University of California, Irvine.

Although the reviewers listed above have provided many constructive comments and suggestions, they were not asked to endorse the conclusions or recommendations, nor did they see the final draft of the report before its release. The review of this report was overseen by Katharine

Frase, IBM. Appointed by the National Research Council, she was responsible for making certain that an independent examination of this report was carried out in accordance with institutional procedures and that all review comments were carefully considered. Responsibility for the final content of this report rests entirely with the authoring committee and the institution.

Contents

SUMMARY   1
    Relevance of Information Technology and Computer Science to Sustainability, 2
    The Value of the Computer Science Approach to Problem Solving, 5
        Systems—Scale, Heterogeneity, Interconnection, Optimization, and Human Interaction, 5
        Iteration, 6
    Computer Science Research Areas, 7
    Strategy and Pragmatic Approaches, 9
        Emphasize Bottom-Up Approaches and Concreteness, 9
        Use Appropriate Evaluation Criteria for Proposals and Results, 9
        Apply CS Philosophy and Approach, 10
        Foster Sustainability Research Through Funding Initiatives, 10
        Foster Needed Multidisciplinary Approaches, 11
        Blend Sustainability and Education, 12

1   ROLES AND OPPORTUNITIES FOR INFORMATION   13
    TECHNOLOGY IN MEETING SUSTAINABILITY
    CHALLENGES
    Opportunities to Achieve Significant Sustainability
        Objectives, 17
      Built Infrastructure and Systems, 18
      Ecosystems and the Environment, 20
      Sociotechnical Systems, 21
    Illustrative Examples in Information Technology and
        Sustainability, 22
      Toward a Smarter Electric Grid, 23
      Sustainable Food Systems, 36
      Sustainable and Resilient Infrastructures, 44
    Conclusion, 50

2   ELEMENTS OF A COMPUTER SCIENCE RESEARCH   51
    AGENDA FOR SUSTAINABILITY
    Measurement and Instrumentation, 55
      Coping with Self-Defining Physical Information, 57
      The Design and Capacity Planning of Physical
          Information Services, 59
      Software Stacks for Physical Infrastructures, 60
    Information-Intensive Systems, 61
      Big Data, 62
      Heterogeneity of Data, 63
      Coping with the Need for Data Proxies, 64
      Coping with Biased, Noisy Data, 65
      Coping with Multisource Data Streams, 66
    Analysis, Modeling, Simulation, and Optimization, 70
      Developing and Using Multiscale Models, 70
      Combining Statistical and Mechanistic Models, 71
      Decision Making Under Uncertainty, 72
    Human-Centered Systems, 77
      Supporting Deliberation, Civic Engagement, Education,
          and Community Action, 79
      Design for Sustainability, 81
      Human Understanding of Sensing, Modeling, and
          Simulation, 82
      Tools to Help Organizations and Individuals Engage
          in More Sustainable Behavior, 82
      Mitigation, Adaption, and Disaster Response, 83
      Using Information from Resource-Usage Sensing, 83
    Conclusion, 85

3   PROGRAMMATIC AND INSTITUTIONAL   86
    OPPORTUNITIES TO ENHANCE COMPUTER
    SCIENCE RESEARCH FOR SUSTAINABILITY
    Computer Science Approaches for Addressing
          Sustainability, 87
    Toward Universality, 93
    Education and Programmatics, 96
    Evaluation, Viability, and Impact Analysis, 100
    Conclusion, 103

APPENDIXES

A   Summary of a Workshop on Innovation in Computing and   107
    Information Technology for Sustainability
B   Biographies of Committee Members and Staff   149

Summary

A broad and growing literature describes the deep and multidisciplinary nature of the sustainability challenges faced by the United States and the world. Despite the profound technical challenges involved, sustainability is not, at its root, a technical problem, nor will merely technical solutions be sufficient. Instead, deep economic, political, and cultural adjustments will ultimately be required, along with a major, long-term commitment in each sphere to deploy the requisite technical solutions at scale. Nevertheless, technological advances and enablers have a clear role in supporting such change, and information technology (IT)[1] is a natural bridge between technical and social solutions because it can offer improved communication and transparency for fostering the necessary economic, political, and cultural adjustments. Moreover, IT is at the heart of nearly every large-scale socioeconomic system—including systems for finance, manufacturing, and the generation and distribution of energy—and so sustainability-focused changes in those systems are inextricably linked with advances in IT. In short, innovation in IT will play a vital role if the nation and the world are to achieve a more sustainable future.

Although the greening of IT—for example, the reduction of electronic waste or of the energy consumed by computers—is an important goal of the computing community and the IT industry, the focus of this report is

---

[1]"Information technology" is defined broadly here to include both computing and communications capabilities.

"greening through IT," that is, the application of computing to promote sustainability broadly.

The aim of this report is twofold: to shine a spotlight on areas where IT innovation and computer science (CS)[2] research can help, and to urge the computing research community to bring its approaches and methodologies to bear on these pressing global challenges. The focus is on addressing medium- and long-term challenges in a way that would have significant, measurable impact.

The findings and recommended principles of the Committee on Computing Research for Environmental and Societal Sustainability concern four areas: (1) the relevance of IT and CS to sustainability; (2) the value of the CS approach to problem solving, particularly as it pertains to sustainability challenges; (3) key CS research areas; and (4) strategy and pragmatic approaches for CS research on sustainability.

RELEVANCE OF INFORMATION TECHNOLOGY AND COMPUTER SCIENCE TO SUSTAINABILITY

An often-cited definition of "sustainability" comes from *Our Common Future*, the report of the Brundtland Commission of the United Nations (UN): "[S]ustainable development is development that meets the needs of the present without compromising the ability of future generations to meet their own needs."[3] The UN expanded this definition at the 2005 World Summit to incorporate three pillars of sustainability: its social, environmental, and economic aspects.[4] This report takes a similarly broad view of the term. Although much of the focus in sustainability has been on mitigating climate change, with efforts aimed at managing the carbon dioxide cycle and increasing sustainable energy sources, there are other important sustainability challenges (such as water management, improved urban planning, supporting biodiversity, and food production) that can also be transformed by advances in computing research and are thus considered in this report.

It is natural when viewing sustainability through the lens of computer science to take a systems view. An elaboration on the broad definition of

---

[2] "Computer science" is defined broadly here to include computer and information science and engineering.

[3] United Nations General Assembly, March 20, 1987, *Report of the World Commission on Environment and Development: Our Common Future*; transmitted to the General Assembly as an Annex to document A/42/427—Development and International Co-operation: Environment; Our Common Future, Chapter 2: Towards Sustainable Development; Paragraph 1, United Nations General Assembly. Available at http://www.un-documents.net/ocf-02.htm.

[4] United Nations General Assembly, 2005 World Summit Outcome, Resolution A/60/1, adopted by the General Assembly on September 15, 2005.

sustainability above is that a system is not sustainable unless it can operate indefinitely into the future. For a system to do so inevitably requires optimization over time and space—goals that are central to much of computer science.

The report *SMART 2020: Enabling the Low Carbon Economy in the Information Age*[5] usefully groups opportunities for applying IT to sustainability into three broad areas: (1) built infrastructure and systems, (2) ecosystems services and the environment, and (3) sociotechnical systems. The following describes each of these areas and outlines applications of IT and opportunities for computer science research:

- *Built infrastructure and systems.* This area includes buildings, transportation systems (personal, public, and commercial), and consumed goods (commodities, utilities, and foodstuffs). IT contributes to sustainable solutions in built infrastructure in numerous ways, from improved sensor technologies (e.g., in embedded sensors in smart buildings) and improved system models, to improved control and optimization (e.g., of logistics and smart electric grids), to improved communications and human-computer interfaces (enabling people to make more effective decisions).
- *Ecosystems and the environment.* This area encompasses assessing, understanding, and positively affecting (or not affecting) the environment and particular ecosystems—these efforts represent crosscutting challenges for many sustainability efforts. The scale and scope of efforts in this area range from local and regional efforts examining species habitats, to watershed management, to understanding the impacts of global climate change. The range of challenges itself poses a problem: how best to assess the relative importance of various sustainability activities with an eye toward significant impact. Additionally, computational techniques will be valuable for developing scientific knowledge and engineering technologies, including improved methods for data-driven science, modeling, and simulation to improve the degree of scientific understanding in ecology.
- *Sociotechnical systems.* Sociotechnical systems encompass society, organizations, and individuals, and their behavior as well as the technological infrastructure that they use. Large and long-lived impacts on sustainability will require enabling, encouraging, and sustaining changes in behavior—on the part of individuals, organizations, and nation-states over the long term. IT, and in particular real-time information and tools, can better equip individuals and organizations to make daily, ongo-

---

[5] The Climate Group, *SMART 2020: Enabling the Low Carbon Economy in the Information Age* (2008). Available at http://www.smart2020.org/publications/.

ing, and significant changes in response to a constantly evolving set of circumstances.

There are, of course, many points of intersection across these areas. For example, eco-feedback devices within the home (a sociotechnical system) interact with the larger, smart grid system (part of the built infrastructure); personal mobile devices (relying on built infrastructure and deployed in a sociotechnical context) provide data that feed into more robust modeling (a crosscutting methodology itself); and so on. In addition, better information about what is happening at an individual or local level can inform broader policy making and decision making.

Smarter energy grids, sustainable agriculture, and resilient infrastructure provide three concrete and important examples of the potential role of IT innovation and CS research in sustainability.

- Moving toward smarter and more sustainable ways of providing electricity will require a rethinking of many aspects of society, including the fundamental electric grid. A forward-looking, sustainable grid scenario presents a fundamentally more cooperative interaction between demand and supply, as well as greater transparency throughout the energy supply chain, with the goal of achieving both deep reductions in peak demand and reductions in overall demand as well as deep penetration of renewables in the supply blend. Information and data management with regard to both time (demand, availability, and so on) and space are essential to making progress toward a smarter, more sustainable electric grid. Computer science research and methodological approaches (in areas including user interfaces and improved modeling and analytical tools) will be needed at all levels to address the broad systems challenges presented by the smart grid.
- With respect to agriculture, there is growing concern regarding whether agricultural productivity can keep pace with human needs. A sustainable food system will be key to ensuring that the world's population receives necessary nutrition without additional damage being done to the environment and society. As with the electric grid, it is in the systems issues in sustainable agriculture that the opportunities for IT seem most salient. Approaches to a sustainable food system include taking a systems view of the challenge; developing methods for measuring the costs, benefits, and impacts of different agricultural systems; assisting in the use of precision agriculture to minimize needed inputs; making information accessible for informed consumption; and developing social networks for local food sourcing. As with the smart electric grid, information and data management are essential to making progress toward a smarter, more sustainable, global food system. Computer science research

and methodological approaches will be needed to address the broad systems challenges—encompassing the environment and ecosystems, social and economic factors, and personal and organizational behaviors affecting food production, distribution, and consumption.

- The development of sustainable and resilient infrastructures poses crosscutting challenges, especially when a broad view of sustainability is taken that encompasses economic and social issues. Contributing to the challenges of increasing the resilience of societal and physical infrastructures is the growing risk of natural and human-made disasters. Enhancing society's resilience and ability to cope with inevitable disasters will contribute to sustainability. Even apart from climate change and resource consumption, the sheer magnitude of Earth's population means that crises, when they happen, will be at scale. Sustainability challenges in this area involve planning and modeling infrastructure, and the anticipation of and response to disasters and the ways in which information technology can assist with developing sustainable and resilient infrastructures.

Sustainability, of course, encompasses much more than the areas and examples outlined above, which are used here to illustrate the breadth of the challenges that need to be faced and the role that computer science and information technology can play.

THE VALUE OF THE COMPUTER SCIENCE APPROACH TO PROBLEM SOLVING

As the sections below discuss, several key underlying philosophical and methodological approaches of computer science are well matched to key characteristics of sustainability problems.

Systems—Scale, Heterogeneity, Interconnection, Optimization, and Human Interaction

Sustainability problems often share challenges of scale—sometimes due to the size of the problem space (e.g., geographic or planetary scale), sometimes due to the potential range of impact (e.g., widespread potential health or economic impacts), and often due to both. Sustainability problems are also typically heterogeneous in nature—there is almost never just one variable contributing to the challenge, or one avenue to a solution. Inputs, solutions, and technologies that can be brought to bear on any given problem vary a great deal. Most sustainability challenges emerge in part due to interconnection—multiple interlocking pieces of a system all having effects (some expected, some not) on other pieces of the system. Solutions to sustainability challenges typically involve finding near-optimal trade-offs among competing goals, typically under high degrees of uncertainty in both the systems and the goals. Hence, methods for finding robust solutions are critical. And finally, human interaction with systems can play a role in both developing solutions and contributing to challenges.[6]

In addition to systems challenges, many sustainability challenges, particularly those related to infrastructure such as smarter transportation or electric systems, involve architecture. Architecture encompasses not just structural connections among subsystems, but also expectations regarding what a system will do, how it will perform, what behaviors are within bounds, and how subsystems (or external actors) should interact with the system as a whole. A system's architecture instantiates early design decisions and has a significant effect on the uses, behaviors, and effects of the system over its life cycle long past the time when those decisions were made. As a result, larger-scale systems of necessity merit significant attention and resources devoted to architecture. As computer and information systems have become global in scale, the disciplines of computer science and software engineering have grappled with the challenges of architecture as they pertain to large-scale systems working over large geographic areas with countless inputs and millions of users. Lessons from architecting hardware, software, networks, and information systems thus have broader applicability to the processes of the structuring, designing, maintaining, updating, and evolving of infrastructure in pursuit of sustainability.

FINDING: Although sustainability covers a broad range of domains, most sustainability issues share challenges of architecture, scale, heterogeneity, interconnection, optimization, and human interaction with systems, each of which is also a problem central to CS research.

Iteration

Given the scope and scale of many of the sustainability challenges faced today, it is very likely that no one solution or approach will suffice, even for those challenges that are comparatively easy to state (such as, "Reduce greenhouse gas emissions"). Thus, multiple approaches from multiple angles will need to be tried. Moreover, the urgency of acting in the face of threats to biodiversity and consequences of global climate change means that the best-known options need to be deployed quickly

---

[6]Of course, many other scientific disciplines offer useful methodological approaches to sustainability, some of which overlap with what computer science offers. This report focuses on computer science, as directed in the study committee's statement of task (see the Preface).

while the adaptive redesign of the deployed system continues to be supported as advances in scientific understanding, changes in technology, and evolution in political and economic systems are incorporated. Thus iteration—adjusting, refining, and learning from ongoing efforts—will be essential, and it will often have to be done at a societal and planetary scale.

Iteration is another core strength of computer science, and learning from iterative approaches to large-scale software systems and applications, and large-scale software engineering and system deployment generally, can help with large-scale sustainability challenges. The approach has been demonstrated in such specific applications as the engineering of the global Internet and the deployment of web search and has been used effectively in a wide array of successful software engineering projects.

Because sustainability challenges involve complex, interacting systems of systems undergoing constant change, a data-driven, iterative approach will be essential to making progress and to making needed adjustments as situations change. One approach is to deploy technology in the field, using reasonably well-understood techniques, at first to explore the space and map gaps that need work. Data and models developed on the basis of this initial foray can then help provide context for developing qualitatively new techniques and technologies for contributing to even better solutions.

> **FINDING: Fast-moving iterative, incrementally evolving approaches to problem solving in computer science, which were critical to building the Internet and web search engines, will be useful in solving sustainability challenges.**

COMPUTER SCIENCE RESEARCH AREAS

Despite numerous opportunities to apply well-understood technologies and techniques to sustainability, there are also hard problems—for example, the mitigation of climate change—for which current methods offer at best partial solutions and the pressing nature of the challenges motivates rapid innovation. This section describes some salient technical research areas and outlines a broad research agenda for CS and sustainability.

> **FINDING: Although current technologies can and should be put to immediate use, CS research and IT innovation will be critical to meeting sustainability challenges. Effectively realizing the potential of CS to address sustainability challenges will require sustained and appropriately structured and tailored investments in CS research.**

The committee selected four broad CS/IT research areas meriting attention in order to help meet sustainability challenges—all of which contain elements of sensing, modeling, and action. The following list is not prioritized. Efforts in all of these areas will be needed, often in tandem.

- Measurement and instrumentation;
- Information-intensive systems;
- Modeling, simulation, and optimization; and
- Human-centered systems.

The areas correspond well to measurement, data mining, modeling, control, and human-computer interaction, which are well-established research areas in computer science. This overlap of selected research areas with established research areas has positive implications: research communities are already established, and it will not be necessary to develop entirely new areas of investigation in order to effectively address global sustainability challenges. Nonetheless, finding a way to achieve that impact may require new approaches to these problems and almost certainly new ways of conducting and managing research.

The ultimate goal of much of computer science in sustainability can be viewed as informing, supporting, facilitating, and sometimes automating decision making that leads to actions with significant impacts on achieving sustainability objectives. The committee uses the term "decision making" in a broad sense—encompassing individual behaviors, organizational activities, and policy making. Informed decisions and their associated actions are at the root of all of these activities.

> **FINDING: Enabling and informing actions and decision making by both machines and humans are key components of what CS and IT contribute to sustainability objectives, and they demand advances in a number of topics related to human-computer interaction. Such topics include the presentation of complex and uncertain information in useful, actionable ways; the improvement of interfaces for interacting with very complex systems; and ongoing advances in understanding how such systems interact with individuals, organizations, and existing practices.**

> **PRINCIPLE: A CS research agenda to address sustainability should incorporate sustained effort in measurement and instrumentation; information-intensive systems; analysis, modeling, simulation, and optimization; and human-centered systems.**

STRATEGY AND PRAGMATIC APPROACHES

For computer science to play an effective part in meeting global sustainability challenges, priority should be given to research that addresses one or more important sustainability challenges and that offers the prospect of tangible impact, either directly or through game-changing contributions that offer leveraging opportunities for other domains. The research areas listed in the section above are the committee's recommended starting place.

An ongoing challenge is for IT experts and CS researchers to ensure that technologies and approaches represent usable and appropriate solutions, that they are highly effective, and that they take advantage of the deepest and most powerful insights that can be brought to bear.

Emphasize Bottom-Up Approaches and Concreteness

The committee believes that CS research on sustainability is generally best approached not by striving for universality from the start, but instead by beginning from the bottom up: that is, by developing well-structured solutions to particular, critical problems in sustainability, and later seeking to generalize these solutions. Indeed, this has been a fruitful approach in many other application areas. Progress in many needed advances will require CS research (as described earlier), but those advances may not be immediately evident as universal approaches. Rather, to be judged as a significant contribution at the intersection of CS research and sustainability, the contribution must first have the potential to make a real difference in moving toward a more sustainable future. Embracing the concrete will help researchers hone and filter their approaches, and multiple and adapted applications will emerge. Many potential new applications are developed and find their ultimate universality through bottom-up cycles of change and through the iterative process of design that promotes those cycles of change. Past successful examples of this approach include Internet protocols, machine learning, object-oriented languages, and databases.

Use Appropriate Evaluation Criteria for Proposals and Results

A premature focus on universality would be damaging to high-impact sustainability solutions. However, to be considered successful, CS research on sustainability must ultimately contribute to generalizable knowledge about sustainability, and the contribution or proposed solution should, at the same time, require new computational techniques or thinking beyond the current state of the art in computing. Establishing metrics for multidisciplinary work that are both actionable and meaningful across participating disciplines is challenging, and the specific criteria for judging research success should evolve over time, with members of the community proposing and debating what constitutes the most worthy research. The committee emphasizes, however, the criterion of having the potential to make a real difference—that is, to make significant progress on social, economic, and environmental sustainability challenges.

> **PRINCIPLE: There should be strong incentives at all stages of research for focusing on solving real problems whose solution can make a substantial contribution to sustainability challenges, along with in-depth metrics and evaluative criteria to assess progress.**

Apply CS Philosophy and Approach

The solutions for real problems referred to in the principle above should be designed such that they embed the best of CS design and systems learnings—modularity, isolation, simplicity, and so on. Then researchers and practitioners should experiment with, apply, and pilot solutions to specific problems, looking for the successes and reapplying and adapting them to other applications and developing universality, while building the applicability and impact. Such work will need to be done across disciplinary boundaries and involve experts from many fields. Just as specific proposed solutions will need to be assessed in an iterative fashion, so too the research enterprise will need to have informed checkpoints and evaluative criteria in order to ensure that the goal of having a real impact is being met. Thus the committee urges an emphasis on interdisciplinarity, iteration, and high-level information sharing to assess progress.

Foster Sustainability Research Through Funding Initiatives

Programmatically, traditional computer science research funding approaches are unlikely to be adequate to address the need discussed here. The National Science Foundation (NSF) is a primary funder of research in computer science in the United States. The former Information Technology Research programs at NSF and the current Cyber-enabled Discovery and Innovation Program are good examples of multidisciplinary programs, demonstrating that such efforts are feasible. But such programs are still a small minority among funding programs, and in the committee's view most review panels on most of the programs related to CS research are not generally favorable toward funding domain-specific projects. The committee is encouraged by the establishment of Science, Engineering, and Education for Sustainability (SEES) as an NSF-wide area of investment. SEES aims for a systems-based approach to "advance science, engineering, and education to inform the societal actions needed for environmental and economic sustainability and sustainable human well-being"[7] and places an emphasis on interdisciplinary efforts. It provides a programmatic opportunity to put the recommended principles of this report into practice at NSF. For the field of computer science, efforts such as this can serve as a model for conceptualizing funding structures in order to take the greatest advantage of the depth of IT and CS innovation that the core discipline can offer to the rich and globally important problem space of sustainability.

Foster Needed Multidisciplinary Approaches

The type of work described above will have to be done across disciplinary boundaries and to involve experts from many disciplines, as well as individuals who themselves have deep expertise in more than one discipline. Among the several opportunities for enhancing multidisciplinary approaches are scholarships that emphasize the development of expertise in complementary disciplines, and regular, high-level summits involving CS and sustainability experts—practitioners and researchers—to inform shared research design, assess progress, and identify gaps and opportunities.

Research institutions—both universities and funding organizations—could better address the needs of authentic multidisciplinary research, in terms of adjustments to how individuals are evaluated and in terms of publications, funding, criteria for promotion, infrastructure for sustained collaboration, and cross training.

> **PRINCIPLE: Encourage research at and across disciplinary boundaries, well informed by specifics and well structured to handle scale, data, integration, architecture, simulation, optimization, iteration, and human and systems aspects. CS research in sustainability should be an interdisciplinary effort, with experts in the various fields of sustainability being equal partners in the research.**
>
> **PRINCIPLE: Refine funding and programmatic options to reinforce and provide incentives for the necessary boundary crossing and integration in CS research to address sustainability challenges. In particular, funding, promotion, and review and assessment (peer**

---

[7] SEES mission statement. Available at http://www.nsf.gov/funding/pgm_summ.jsp?pims_id=504707.

review) models should emphasize in-depth integration with data and deployments from the constituent domains.

Blend Sustainability and Education

A shifting of the culture of CS to embrace sustainability more fully as an important and fruitful application area for research needs to include educating CS students about ways to have an impact with computing, computation, and systems approaches in important areas. Such a shift in culture would encourage students to develop domain expertise and to collaborate directly with domain experts while in graduate school or in preparing for graduate work. Such a shift also requires a culture of experimentation and innovation in the application of computer science.

Adjusting education within the target domains is as important as shifting the culture in CS. Information and data are critical to understanding the challenges, to formulating and deploying solutions, to communicating results, and to facilitating learning and new behaviors based on the results of the work. Thus a significant component of meeting virtually all sustainability challenges is to infuse computational thinking and approaches that are rich in CS and IT into the deploying industry and agencies. This component needs to include cross training students in multiple fields to create "champions" who can bring a CS perspective into other arenas. Sustainability is a challenge that will persist for generations; sustained commitment will be necessary, as well as continuing innovation in support of efforts to meet sustainability challenges.

PRINCIPLE: Undergraduate and graduate education in computer science should provide experience in working across disciplinary boundaries. Graduate training grants and postdoctoral fellowships should support training in multiple disciplines. Undergraduate and graduate programs should include tracks that offer introductory and intermediate course work in such sustainability areas as life-cycle analysis, agriculture, ecology, natural resource management, economics, and urban planning.

1

Roles and Opportunities for Information Technology in Meeting Sustainability Challenges

Innovation in computing, information, and communications technology is at the heart of nearly every large-scale socioeconomic system. Computing underlies and enables systems that affect our lives every day—from financial and health systems to manufacturing, transportation, and energy infrastructures. One important consequence is that advances in computing are critical enablers of change for addressing the growing sustainability challenges facing the United States and the world. A key finding of this report is that information technology (IT)[1] will play a vital role in achieving a more sustainable future and that research and innovation in computing, information, and communications technologies are consequently critical to addressing the broad range of sustainability challenges (Box 1.1).

The critical global challenges in sustainability are deep, and solutions will require input from many disciplines. Fortunately, there are numerous opportunities to apply IT innovations in ways that will have a profound influence on sustainability efforts across many areas, including the ecological and environmental sciences, numerous engineering fields, public policy and administration, and many other areas. The National Research Council's (NRC's) Committee on Computing Research for Environmental and Societal Sustainability is aware that there is significant effort aimed at making IT itself "greener" and recognizes that these efforts are important.

---

[1]The committee uses the familiar acronym "IT" (information technology) to encompass computing, information, and communications technologies broadly.

> **BOX 1.1**
> **A Note on the Definition of "Sustainability" and the Focus of the Committee**
>
> An often-cited definition of "sustainability" comes from the Brundtland Commission of the United Nations (UN): "[S]ustainable development is development that meets the needs of the present without compromising the ability of future generations to meet their own needs."[1] The UN expanded this definition at the 2005 world summit to incorporate three pillars of sustainability: its social, environmental, and economic aspects.[2] This report takes a similarly broad view of the term. Although much focus in sustainability has been on mitigating climate change, with efforts aimed at managing the carbon dioxide cycle and increasing sustainable energy sources, the committee recognizes that there are numerous additional sustainability challenges that could be assisted by advances in computing and information technology and computing[3] research. The committee's focus is on addressing medium- and long-term challenges in a way that has significant and ideally, measurable, impact.
>
> ---
>
> [1]United Nations General Assembly (March 20, 1987). *Report of the World Commission on Environment and Development: Our Common Future;* transmitted to the General Assembly as an Annex to document A/42/427—Development and International Co-operation: Environment; Our Common Future, Chapter 2: Towards Sustainable Development; Paragraph 1. United Nations General Assembly. Available at http://www.un-documents.net/ocf-02.htm.
> [2]United Nations General Assembly, 2005 World Summit Outcome, Resolution A/60/1, adopted by the General Assembly on September 15, 2005.
> [3]The term "computing" is used generally in this report and is meant to encompass information and communications technologies (ICTs). Thus "computing" and "ICTs" are used interchangeably throughout the report.

The greening of IT, through efforts such as reducing data-center energy consumption and electronic waste, should be and is an important goal of the computing community and IT industry.[2] However, the focus of this report is on what could be termed "greening through IT," the use of

---

[2]The 2010 OECD report "Greener and Smarter: ICTs, the Environment and Climate Change" (in OECD, *OECD Information Technology Outlook 2010,* OECD Publishing) notes that impacts from ICT life cycles (including not just use but also production and end of life) need to be considered in order to understand complete impacts. A recent *McKinsey Quarterly* article, "Clouds, Big Data, and Smart Assets: Ten Tech-Enabled Business Trends to Watch," by Jacques Bughin, Michael Chui, and James Manyika, offered some cause for optimism regarding green IT: "Electricity produced to power the world's data centers generates greenhouse gases on the scale of countries such as Argentina or the Netherlands, and these emissions could increase fourfold by 2020. McKinsey research has shown, however, that the use of IT in areas such as smart power grids, efficient buildings, and better logistics planning could eliminate five times the carbon emissions that the IT industry produces." *McKinsey Quarterly* 5(3):1-14.

computing and IT across disciplines to promote sustainability in areas and systems in which advances in information and communications technology (ICT) could have significant positive impact.[3]

The committee believes that some of the most profound fundamentals within the field itself are suggestive of the unique contributions that computer science (CS) and ICTs can make to sustainability. For instance, the very notion of automated "query-able" structured data is at the heart of much of computer science. The scope and scale of the sustainability challenge are coupled with vast amounts of relevant data, which makes deep insights into the challenges of collecting, structuring, and understanding those data essential. Computational thinking is critical to solving almost any large problem. The committee's focus is on problems that are intellectually challenging, grounded in IT and CS, and important for sustainability—that is, a kind of "Pasteur's octant." See Figure 1.1.

Despite the profound technical challenges presented by sustainability and the huge potential role for IT and CS, the committee recognizes that sustainability is not, at its root, a technical problem, nor will merely technical solutions be sufficient. Instead, solutions ultimately will require deep economic, political, and cultural adjustments, as well as major, long-term commitment in each sphere in order to put technical advancements and enablers in operation at scale. Nevertheless, technological advances and enablers can be developed and shaped to support such change, while continuing to support enduring human values in the process. Information technology can serve as a bridge between technical and social solutions

---

[3]The community has already begun addressing this challenge. Bill Tomlinson's book *Greening Through IT: Information Technology for Environmental Sustainability* (Cambridge, Mass.: MIT Press, 2010) explores how IT can address sustainability challenges at scale. A 2009 article by Carla Gomes, "Computation Sustainability: Computational Methods for a Sustainable Environment, Economy, and Society" in *The Bridge* 39(4):5-13, provides examples of computational research being applied to domain fields (biodiversity and renewable energy sources). Gomes's work is an important component of computational sustainability; the present report explores the broader potential for research and innovation in CS and IT to have an impact on sustainability. Additionally, the National Science Foundation's Directorate for Computer and Information Science and Engineering and the Computing Community Consortium (CCC) jointly sponsored a workshop on the Role of Information Sciences and Engineering in Sustainability. The full report of the workshop, *Science, Engineering, and Education of Sustainability: The Role of Information Sciences and Engineering*, which discusses research directions for IT as it relates to sustainability, is available at http://cra.org/ccc/docs/RISES_Workshop_Final_Report-5-10-2011.pdf. This report is well aligned, in terms of research areas, with the CCC report. Additionally, the committee concurs with the CCC report Section 4, titled "The Power of Use-Inspired (Collaborative) Fundamental Research." The present report expands on this theme in Chapter 3, especially in regard to the strength of computer science as a discipline and what it can contribute to sustainability objectives.

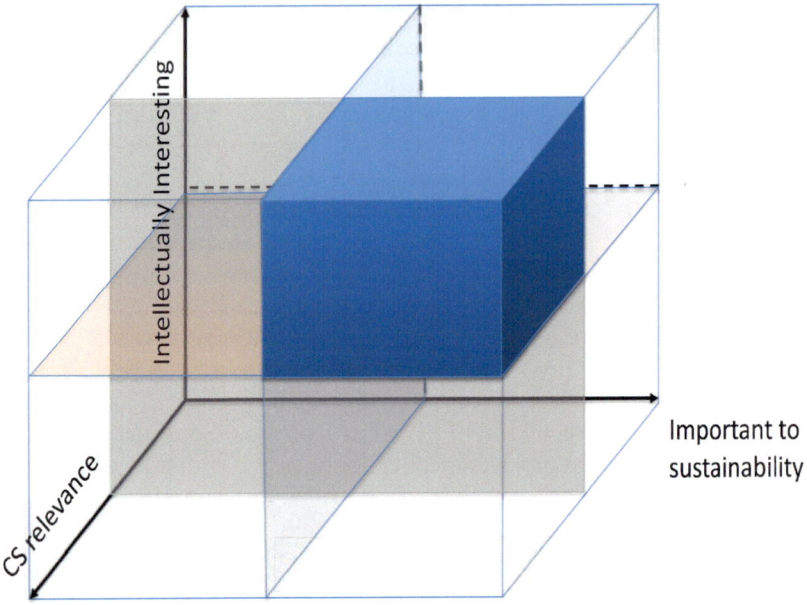

FIGURE 1.1 The committee's focus is on problems at the intersection of significant intellectual merit, relevance to computer science (CS), and importance to sustainability.

by enabling improved communication and transparency for fostering the necessary economic, political, and cultural adjustments.[4]

Furthermore, sustainability problems are typically heterogeneous in nature—there is almost never just one variable contributing to the challenge or one avenue to a solution. Inputs, solutions, and technologies that can be brought to bear on any given problem vary themselves a great deal. Most sustainability challenges emerge in part due to interconnection—a result of multiple interlocking pieces of a system all having effects (some expected, some not) on other pieces of the system. Solutions to sustainability challenges typically involve finding near-optimal trade-offs among competing goals, typically under high degrees of uncertainty in both the systems and the goals.

In addition to noting the crosscutting nature of many sustainability challenges, it is important to recognize the emergent qualities that typify the sorts of systems being discussed here. Some projections of what might

---

[4]E. Ostrom. A general framework for analyzing sustainability of social-ecological systems, *Science* 325:419-422 (2009).

be accomplished with the savvy application of known technologies or near-term research are straightforward, even in systems and domains as complex as these. However, in such complex systems and domains there are likely to be emergent behaviors and properties as well—both toward and away from desired outcomes. IT practitioners have proven remarkably adept at innovating flexibly when previously unanticipated systems behaviors have demanded responses. The complexity and unpredictability of the results of unsustainable human activities require an innovative and flexible approach to solving or mitigating sustainability problems and their impacts, and IT researchers and practitioners are skilled at innovating and developing flexible solutions in dynamic environments. The committee believes that computing researchers and research approaches will be essential to grappling with current and future systems challenges in sustainability.

This report has three chapters. Chapter 1 elaborates on domains of potential impact in order to illustrate the role and the available opportunities of IT on the broader path toward sustainability. It address the question, In what ways and where can computing research have measurable, significant impact? Chapter 2 describes methods and approaches in discussing the questions, How do fundamental research questions and approaches in computing intersect with sustainability challenges, and how can problem solving and research methodologies in computing research and IT innovation be brought to bear on sustainability? In particular, the committee views one important goal of computer science in sustainability as informing, supporting, facilitating, and sometimes automating decision making—decision making that leads to actions that will have significant impacts on achieving sustainability objectives. Aimed primarily at computer science researchers, Chapter 3 articulates why the interplay between addressing sustainability challenges and computer science research merits attention, and how that interplay offers deep and compelling opportunities for progress in multiple dimensions. Appendix A summarizes presentations and discussions at the Workshop on Innovation in Computing and Information Technology for Sustainability, organized by the committee. Biographies of the committee members are presented in Appendix B.

OPPORTUNITIES TO ACHIEVE SIGNIFICANT SUSTAINABILITY OBJECTIVES

Forward-looking IT innovations and sustained research can have significant positive impact for sustainability across many areas. For the purposes of this report, the areas are clustered as follows: built infrastructure and systems, ecosystems services and the environment, and sociotechnical systems.[5] Each of these is described briefly below. There are obvious multiple intersection points in these three distinct areas of opportunity. For example, eco-feedback devices (tools that provide instant information on environmental impact) within the home, a sociotechnical system,[6] interact with the larger smart grid system, part of the built infrastructure; personal mobile devices, relying on built infrastructure and deployed in a sociotechnical context, provide data that feed into more robust modeling, a crosscutting methodology, and so on. In all of these domains, as potential solutions are deployed, careful attention will need to be paid to iterate over and evaluate solutions to ensure that progress made in one dimension of a given sustainability problem is not later offset by an unanticipated outcome or side effect in another dimension. The next major section, "Illustrative Examples in Information Technology and Sustainability," provides crosscutting examples of domains in which IT can support and strengthen sustainability efforts.

Built Infrastructure and Systems

Built infrastructure and systems include buildings (residential and commercial), transportation systems (personal, public, and commercial), and consumed goods (commodities, utilities, and foodstuffs). The Climate Group's *SMART 2020* report examined the use of information and communication technology in built infrastructure in several key areas, including smart buildings, smart logistics, and smart electric grids. According to that report, these three areas alone provide a potential reduction in greenhouse gas (GHG) emissions of 15 percent of global "business as usual" emissions in 2020.[7]

Buildings account for up to 40 percent of energy use in industrialized countries and 40 percent of GHG emissions; in the United States they consume more than 70 percent of the electricity produced.[8] Smart buildings use IT systems to make better use of energy while maintaining indoor health and comfort. The embedded IT monitors and controls environ-

---

[5]Other clusterings are of course possible. The choice of these three was inspired in part by Global e-Sustainability Initiative, *SMART 2020: Enabling the Low Carbon Economy in the Information Age* (2008). Available at http://www.smart2020.org/publications/.

[6]"Sociotechnical systems" encompass society, organizations, and individuals, and their behavior as well as the technological infrastructure that they use.

[7]Global e-Sustainability Initiative, *SMART 2020: Enabling the Low Carbon Economy in the Information Age* (2008). Available at http://www.smart2020.org/publications/.

[8]World Business Council for Sustainable Development, *Energy Efficiency in Buildings: Facts and Trends—Full Report* (2008). Available at http://www.wbcsd.org/pages/edocument/edocumentdetails.aspx?id=13559&nosearchcontextkey=true. See also http://www.eesi.org/buildings.

mental and electrical systems in the building by means of computerized, intelligent networks of sensors and electronic devices.[9] According to the *SMART 2020* report, smart buildings could reduce carbon dioxide emissions by an estimated 15 percent in 2020.[10] The sustainability of structures generally goes well beyond energy, and involves the reuse and recycling of materials, sustainable construction processes, improved indoor air quality, effective water use, and so on.[11]

Smart logistics use IT for more effective supply chains (those dealing with journey and load planning and with personal transportation), both in daily operational use and in long-term planning. Examples of IT contributions include better geographic information systems and design software to promote more effective transport networks, collaborative multi-institutional planning tools to lower the logistical demands associated with desired lifestyles, and better inventory-management tools. Computing innovation can also lead to better management of consumed resources. Smart electric grids use IT tools throughout the power networks to enable optimization. (Potential smart grid applications are described in greater detail in the section "Toward a Smarter Electric Grid," below.)

In addition to reductions that can be achieved in energy consumption, smarter water- and sewage-management systems in the built infrastructure can decrease water consumption and waste. Furthermore, large-scale agriculture necessitates water and supply-chain management; advanced IT can enhance precision agriculture, including the incorporation of technologies to predict crop yields more accurately.[12] (See the section "Sustainable Food Systems," below, for more on food systems broadly.)

Transportation and city and regional planning also provide important opportunities for more sustainable development; computation and IT will be needed to enable significantly more complex planning for the optimizing of investment in new infrastructure. And, changes to manufacturing itself (which incorporates logistics, sensing, transportation, and manipulation) can help with sustainability goals by reducing environmental impacts, conserving energy and resources, and improving safety

---

[9]National Research Council, *Achieving High-Performance Federal Facilities: Strategies and Approaches for Transformational Change,* Washington, D.C.: The National Academies Press (2011).

[10]Global e-Sustainability Initiative, *SMART 2020: Enabling the Low Carbon Economy in the Information Age* (2008). Available at http://www.smart2020.org/publications/.

[11]For an introduction to some of the issues related to achieving high-performance "green" buildings, see National Research Council, *Achieving High-Performance Federal Facilities: Strategies and Approaches for Transformational Change,* Washington, D.C.: The National Academies Press (2011).

[12]National Research Council, *Toward Sustainable Agricultural Systems in the 21st Century,* Washington, D.C.: The National Academies Press (2010).

for the individuals and communities affected by it. IT has a central role in these efforts.

Ecosystems and the Environment

Assessing, understanding, and positively affecting (or not affecting) the environment and particular ecosystems are crosscutting challenges for many sustainability efforts.[13] The scale and scope of such efforts range from local and regional activities examining species habitats, to watershed management, to efforts to increase understanding of the impacts of global climate change. The range of challenges itself poses a problem: how best to assess the relative importance of various sustainability activities with an eye toward significant impact. Nonetheless, in virtually every activity related to meeting sustainability challenges, a critical role is required of data, information, and computation.

Climate science, for example, has been able to take huge leaps forward due to advances in computing research.[14] Computational modeling and simulation of Earth, the atmosphere, oceans, and biota and of their many interactions have long been at the heart of understanding how changes in carbon cycles and hydrological cycles give rise to global climate change and the estimating of future impacts. Sensing, data management, and model formation connect these computational analyses to a vast body of empirical observation and to one another. Such tools allow for the continual improvement of fidelity and can help improve the basic understanding of ows of carbon, nitrogen, and other emissions of interest. These tools also improve the understanding of water and resource usage, of species distributions and biodiversity, and of ways in which human activity perturbs these. Analyses of environmental and ecosystem responses to disturbances (those from GHGs, fire, invasive species, disease) are important to meeting a range of sustainability objectives. Modeling also plays a crucial role in guiding decision makers, by connecting ecological science and research to ongoing ecosystem policy and management. For

---

[13]A recent National Research Council report "capture[s] some of the current excitement and recent progress in scientific understanding of ecosystems, from the microbial to the global level, while also highlighting how improved understanding can be applied to important policy issues that have broad biodiversity and ecosystem effect." National Research Council, *Twenty-First Century Ecosystems: Managing the Living World Two Centuries after Darwin*, Washington, D.C.: The National Academies Press (2011), p. ix.

[14]D.A. Randall, R.A. Wood, S. Bony, R. Colman, R. Fichefet, J. Fyfe, V. Kattsov, A. Pittman, J. Shukla, J. Srinivasan, R.J. Stouffer, A. Sumi, and K.E. Taylor. *The Physical Science Basis. Contribution of Working Group I to the Fourth Assessment Report of the Intergovernmental Panel on Climate Change*, S. Solomon D. Qin, M. Manning, Z. Chen, M. Marquis, K.B. Averyt, M.Tignor and H.L. Miller (eds.), Cambridge, United Kingdom: Cambridge University Press (2007). Available at http://www.ipcc.ch/publications_and_data/ar4/wg1/en/ch1s1-5-3.html.

example, models that jointly capture the interrelationships of multiple variables and their joint uncertainty can support improved understanding and more robust decision making.

Sociotechnical Systems

Large and long-lived impacts on sustainability will require enabling, encouraging, and sustaining desired human behavior—that of individuals, organizations, municipalities, and nation-states—over the long term. Sociotechnical systems designed to aid in behavioral assistance and reinforcement and to provide information about progress are a critical element for global sustainability efforts. Such systems and associated tools are needed at every scale and can be applied to a range of problems, from enabling effective response in times of acute crisis management, to urban planning, to promoting the understanding of behavioral impacts (sometimes referred to as footprint analysis) on carbon, water, and biodiversity.

Institutional behaviors will need to shift in order to realize continuous, long-term environmental changes. Marketing and public education initiatives are important and can contribute to individual and institutional knowledge on best practices. However, real-time information and tools can better equip individuals and organizations to make daily, ongoing, and significant changes in response to a constantly evolving set of circumstances. Information dashboards accessible to key decision makers are an example of how IT can be used to collect, analyze, curate, and informatively present critical information quickly to those who need it most. For example, if the financial incentives for energy utilities shift from an emphasis on delivering more power more cheaply to an emphasis on improving the GHG emissions efficiency of a given level of service, new information will be needed. Gathering such information will require greater visibility and understanding of the dynamics of customer demand, grid capacity, and supply availability. In addition, each of the stakeholders will need more effective means of communicating needs and trade-offs. Similarly, in order for urban planning to promote, say, the reduction of liquid fuel consumption for personal transportation, the processes of street design, zoning, planting, business development, water and waste management, and public transportation need to be coordinated across multiple governing bodies and constituencies.

Personal devices, most notably sensor-rich smartphones, not only provide information and services to their users, but also can provide scientists and researchers with information that may have been missed by traditional operational networks. Furthermore, citizen scientists are increasingly engaged in scientific problem solving, for example by documenting species locations, air quality, and other indicators.[15] In addition, environmental challenges—those caused by damage to the environment from rising ocean water levels and temperatures or those created by the search for and extraction of materials—can be monitored, assessed, and tracked. Information about environmental challenges can also be disseminated using smarter IT. Further advances in the ability to analyze data collected by a wide array of sources will facilitate a better understanding of how environmental crises begin and how to avoid them in the future.

ILLUSTRATIVE EXAMPLES IN INFORMATION TECHNOLOGY AND SUSTAINABILITY

This section contains three illustrative examples of sustainability-related domains in which IT can have significant impact and in which there is both some current activity as well as prospects for significant progress and impact in the future. This set of examples is not meant to be comprehensive and does not re ect a prioritization. Rather, these examples were chosen to illustrate how IT—both currently understood technologies as well as new ones—could be brought to bear on sustainability challenges and also to show the range and variability of what is meant by sustainability. Each example area listed below cuts across the three broad areas outlined above.

- *The smart grid.* In this first example, the grid is clearly part of built infrastructure, but it also has the potential to affect regional ecosystems dramatically as new sources of renewable energy are brought online (for example, solar facilities deployed in deserts will affect the desert ecosystem). Managing the smart grid, from both the supply and the consumption side (which may not be as easily separable in any event) will require sociotechnical systems, such as data management, for humans and human organizations.
- *Food systems.* This second example also encompasses built environments (including the transportation system), the environment, and ecosystems (in various aspects from macro effects on watersheds to strategies for precision agriculture), and, like the smart grid, it requires sophisticated tools and data management to be most effective.
- *The development of sustainable and resilient infrastructures.* This third example poses crosscutting sustainability challenges, especially when considering a broad view of sustainability that encompasses economic

---

[15]W. Willett, P. Aoki, N. Kumar, S. Subramanian, and A. Woodruff, Common sense community: Scaffolding mobile sensing and analysis for novice users, pp. 301-318 in *Proceedings of the 8th International Conference on Pervasive Computing (Pervasive '10)* (May 2010).

and social issues. These challenges include planning and modeling infrastructure and anticipating and responding to increasingly frequent natural and human-made disasters.

Toward a Smarter Electric Grid

Being able to meet the planet's energy needs in a sustainable fashion is fundamentally interwoven with foundational transformations in the design, deployment, and operation of the world's electric grids. The problem is large and complicated, and the committee's framing in this discussion is for descriptive purposes, and is not meant to be complete, to be prescriptive, or to conflict deliberately with other approaches to characterizing the problem.[16] With regard to the electric grid, most analyses of potential paths to stabilizing GHG concentrations involve three interrelated advances: deep efficiency gains, electrifying the demand, and decarbonizing the supply.[17] As a prime example, the United States currently consumes roughly 100 quadrillion British thermal units (Btu) (about 100 exajoules) of energy per year, with flows from supply to demand as illustrated graphically in Figure 1.2. Roughly half of the energy supply goes into the production of electricity. Of that, the largest share is provided by coal, which has the worst GHG intensity of the supplies and is the cheapest and fastest way to increase supply in developing economies. By contrast, essentially all of the renewable and zero-emission supplies also go into electricity production, but these account for a tiny fraction of the energy mix. Their share must increase substantially in order to

---

[16] For instance, a survey paper developed by IBM Research on the computational challenges of the evolving smart grid is oriented around the challenges of data, grid simulation, and economic dispatch: J. Xiong, E. Acar, B. Agrawal, A. Conn, G. Ditlow, P. Feldmann, U. Finkler, B. Gaucher, A. Gupta, F-L. Heng, J. Kalagnanam, A. Koc, D. Kung, D. Phan, A. Singhee, and B. Smith, *Framework for Large-Scale Modeling and Simulation of Electricity Systems for Planning, Monitoring, and Secure Operations of Next Generation Electricity Grids,* Special Report in Response to Request for Information: Computation Needs for the Next-Generation Electric Grid, DOE/LBNL Prime Contract No. DE-AC02-05CH11231, Subcontract No. 6940385 (2011); M. Ilic, Dynamic monitoring and decision systems for enabling sustainable energy services, *Proceedings of the IEEE* 99:58-79 (2011), notes the fundamental role of a man-made power transmission grid and its IT in enabling sustainable socioecological energy systems. J. Kassakian, R. Schmalensee, K. Afridi, A. Farid, J. Grochow, W. Hogan, H. Jacoby, J. Kirtley, H. Michaels, I. Pérez-Arriaga, D. Perreault, N. Rose, and G. Wilson, *The Future of the Electric Grid: An Interdisciplinary MIT Study,* available at http://web.mit.edu/mitei/research/studies/the-electric-grid-2011.shtml#report, aims to provide an objective description of the grid today and makes recommendations for policy, research, and data for guiding the evolution of the grid.

[17] California Council on Science and Technology, *California's Energy Future: A View to 2050,* Sacramento (2011). Available at http://www.ccst.us/publications/2011/2011energy.pdf.

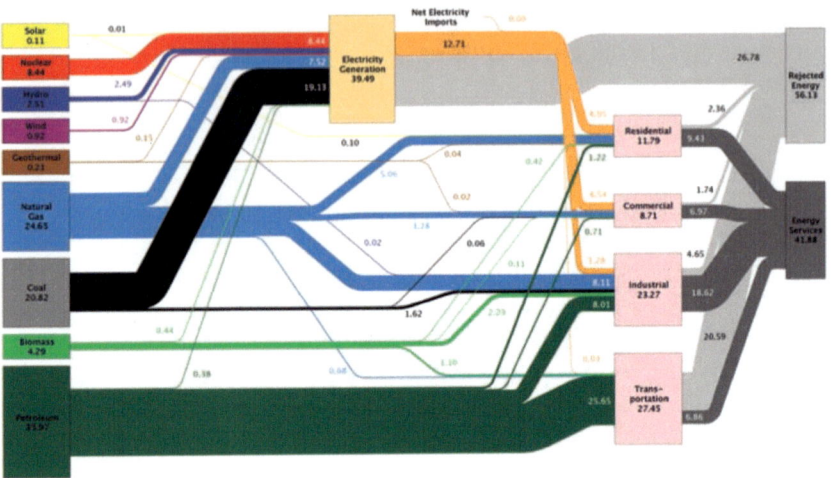

FIGURE 1.2 Current U.S. national energy flow. Roughly half of the 100 quads (10^{15} Btu) is lost, most coal goes to electricity, electricity goes almost equally to residential buildings, commercial buildings, and industrial processes. SOURCE: Lawrence Livermore National Laboratory (2010). Data are based on DOE/EIA-0384 (2009). Available at https://flowcharts.llnl.gov/.

reduce the GHG intensity of the delivered electricity. Doing so will dramatically change the nature of the supply, however, since the availability of these resources varies with natural factors, such as wind and sun, rather than being dispatched as needed to meet demand. Furthermore, the geographic placement of these supplies is governed by natural factors, and so the points at which they attach to the grid, and therefore the pattern of flow from supply to demand and hence the power lines, stations, and devices used to convey these flows of electricity, may be quite different from the flows associated with traditional power plants. This has implications for IT, since the information-management problem for distributed energy production is fundamentally different from that for more centralized production. Managing electricity produced by a half million windmills requires advanced IT—data management, algorithms, and analytics—whereas managing a few hundred coal-fired power plants is a much simpler proposition from an IT perspective.

Already a significant fraction of the supply in the U.S. national energy flow is wasted in the generation, transportation, and conversion of this electrical energy, and of that delivered into residential and commercial buildings and industrial processes, much is wasted through inefficient or ineffective usage. Moreover, reducing the GHG emissions associated with transportation and industrial processes, which are currently dominated by liquid fossil fuels, will involve electrification (e.g., plug-in hybrid or electric vehicles) and hence will further increase demand. Major efficiency gains, the accommodation of variable supplies, and electrification are all likely to involve change in the patterns and practices of the institutions and individuals that represent the demand, which in large part rests on access to actionable information. Innovation in IT and its use underlie all aspects of such a transformation, as described below.

Electric grids can be characterized by their key components: generation, transmission, distribution, and load. Typically, each of these components has been addressed in isolation. Although multiple approaches to transforming electric grids fit within the term "smart grid," the fundamental change in the future will likely be to treat the key components together, as an interrelated system. Whereas other disciplines will contribute primarily to the advance of the physical components comprising the elements of the energy supply chain, IT is expected to govern how these elements behave and how the complex system as a whole functions—that is, what properties it exhibits. The section below first describes several challenges presented by smart grids and then outlines approaches to addressing these challenges, especially from an IT perspective. Finally, a discussion of the specific role of computer science research and innovation in IT is offered.

Challenges for the Modern Electric Grid

Four main challenges for the modern electric grid are discussed below:

- Increased electricity consumption and corresponding growth of the grid;
- The current model of load-following supply, in which capacity is dispatched on the basis of real-time power demand, with coarse predictive analytics deployed to ensure that enough will be available;
- The difficulty in implementing a supply-following model, in which demand is managed to better match the available supply; and
- Appropriate accounting for currently externalized costs.

Increased Consumption Increased productivity and improved standards of living correlate closely with increased energy consumption. Even in the United States, where the energy-to-gross domestic product (GDP) ratio has been steadily improving through technological improvements and efficiency measures, especially since the oil crisis in 1973, overall energy consumption continues to increase. This is an especially serious problem in recently industrialized nations, such as China. Continuing increases in consumption pose multiple challenges.

In terms of generation, rapid increase in electricity production tends to skew the supply blend toward carbon- and particulate-heavy sources such as coal, because such supplies are currently easier to bring online quickly when needed. This trend further compounds the GHG emission problem. Lower-carbon options, such as nuclear power, present other hazards, and renewable sources cannot typically be dispatched on demand, impose other environmental impacts, or are remote from areas of dense consumption. Wasteful production and manufacturing practices, especially in newly industrialized or rapidly growing economies, further compound the climate impact. IT cannot provide generation, but it can enable more effective use of generation facilities to meet increased demand, facilitate the shift toward more desirable supplies, and help manage the increasing demand.

In addition to providing adequate supply to meet growing demand—which clearly cannot continue indefinitely—it must be possible to deliver the generated energy through the transmission grid and distribution tree reliably and safely. Each power line and each piece of electrical equipment has limited capacity and lifetime.[18] Steering actions or switches do not determine the amount of power transferred along each line explicitly, as is done in networks involving transportation, communications, or even water distribution. Instead, the amount of power is determined implicitly, by the underlying physics associated with a distributed collection of loads, connected to a differently distributed collection of generators, through a particular interconnection of wires and transformers. Individual consumers decide independently how much to draw at each load point, and a centralized system operator orchestrates the production at each of the generators in order to match the supply to the demand in real time within the capacity limits of each line and transformer, and within emissions limits set for each generator. This constrained optimization problem is relatively tractable if the transmission and distribution infrastructure is sufficiently overprovisioned. But, as more of its capacity is demanded, the problem becomes substantially more difficult. A network of communicating sensors is overlaid onto the grid to monitor its distributed state, and sophisticated algorithms are used to predict demand, model the ows, schedule generation, and adjust the limited set of control points that are present. Thus, the ability to meet increased demand through the physical

---

[18]For example, as more power is transferred along a line, more heat is generated, causing the line to stretch, become thinner, and sag. This increases the resistance of the line, causing it to heat further, and increases parasitic losses due to capacitance to the ground, which increases demand. All of these factors contribute to failures, which eliminate portions of the transmission or distribution infrastructure and thereby place potentially excessive demand on remaining portions.

infrastructure that exists at any particular time is almost entirely through advances in IT.

These challenges are further complicated by the changing nature of the load and the broader introduction of distributed generation. Unlike purely resistive loads, such as heating elements and incandescent bulbs, complex loads effectively cause a portion of the delivered power (called reactive power) to be returned through the grid to the generator. Historically, such "non-unit power factor" loads were predominantly induction motors, which introduce a fairly simple phase shift in the alternating current (AC) waveform. But switching power supplies, such as those used in computers, uorescent bulbs, battery chargers, and electronics direct current (DC) adapters, introduce complex distortions on the AC waveforms. Residential grid-tie solar installations reverse the ow of electricity within portions of the local distribution tree. And the introduction of electric vehicles potentially introduces high point loads during recharging. Many of these new complex loads already possess communications and computations capabilities, and so they could potentially be a vanguard of using IT to condition demand in order to be "good citizens" of the grid.

Compounding all of these issues still further are the economic structures that impinge on all aspects of generation, delivery, and demand at a range of timescales. On an operational basis, collections of suppliers, consumers, and brokers typically participate in highly volatile wholesale energy markets at various granularities and timescales—a day ahead, an hour ahead, a minute ahead. Meanwhile, consumers typically experience relatively stable retail pricing. Compounding all of these issues further, utilities and the utility supply industry are still largely incentivized to produce and deliver more energy, not less. Economic or other incentives to curb growth are lacking in most parts of the world. A notable exception to this is net-metering—mechanisms that allow electricity consumers to offset their usage of electricity provided by the grid, and thus to lower their cost, by generating their own electricity onsite, typically through rooftop solar photovoltaic installation. Basically, this can be thought of as the meter spinning backward when local generation exceeds local demand. Although net-metering is comparatively common, its penetration is modest enough that it can be incorporated as offsetting demand in the neighborhood distribution tree, without appreciable impact on transmission needs. Broader, less-tangible incentives include the personal satisfaction of obtaining a zero-net lifestyle, potentially opening paths toward the decoupling of quality of life from energy usage. IT has an important role in doing the complex accounting and providing visibility into the consumption and production of otherwise invisible resources.

Current Model of Load-Following Supply Grid operation predominantly involves orchestrating a portfolio of dispatchable supplies, including baseline (nuclear, coal, and hydroelectric power), intermittent (combined-cycle gas turbine), and peaker (simple-cycle natural gas) power plants to supply precisely the real-time power demand across normal variations, spikes, and infrequent peaks in the load. Of course, the demand is not specified explicitly, but implicitly in the use of electricity. Typically, independent service operators perform day-ahead demand prediction for their entire grid, with hour-ahead and even minutes-ahead adjustments, to drive scheduling and market mechanisms while providing adequate generation capacity at all specific points in the transmission grid over time. A certain fraction of online capacity is retained as "spinning reserve" and is used to match short-term changes in demand. An imperfect matching of supply to demand manifests in degraded power quality (such as voltage sags or surges, and frequency variation). Challenges include the following:

• *High cost of peak demand.* Since generation and transmission capabilities must be built out to meet the peak demand, this peak drives overall investment. However, because there is significant variation in demand, a substantial portion of this investment experiences very low utilization. Fundamentally, load following relies on statistical multiplexing of independent loads; even though the individual loads are very bursty, the aggregate of many such loads is relatively smooth and predictable. However, correlations in the loads, such as air conditioning on hot summer afternoons or refrigerator compressor cycles at breaks in Superbowl action, generate very large aggregate peak demand. Means of power generation with short ramp-up times tend to have low efficiency and high GHG emissions and operating costs.

• *Prediction accuracy and market volatility.* A mismatch of predicted and actual demand leads to large and rapid fluctuations in wholesale energy prices. Each new broad-based usage change (for example, the increased uptake of plasma television sets, compact fluorescent lamps, electric vehicles, and so on) raises concerns of prediction accuracy. Paradoxically, by eliminating waste, energy-efficiency measures can lead to larger peak-to-average ratios and potentially lower prediction accuracy, making the grid harder to manage.

• *Storage limitations.* Grid-level storage exists in the form of pumped-storage hydroelectricity, compressed air, thermal energy storage, batteries, and a few other possibilities, but storage capacities remain limited. Storage is typically expensive, and turnaround efficiencies (the energy extracted from storage relative to the amount stored) tend to be low. Small-scale battery storage is prevalent but expensive, and the number of recharge cycles, and hence battery lifetime, is limited. When the effects of manufacturing and the disposal of batteries are taken into account, such storage may have a net negative environmental impact. Midscale storage (say, 1 to 100 kilowatt-hours) is almost non-existent, although ow batteries and electrolysis/fuel cell options remain in development.[19]

- *Non-dispatchable supplies.* Most renewable sources of energy, such as wind, solar, and wave, are non-dispatchable. That is, they are available only at certain times and in magnitudes determined by various environmental factors; they cannot be summoned on demand. Gross features, such as the incident solar radiation over the course of the day or the seasonal patterns in wind, are much more predictable than fine features, such as occlusion due to passing clouds or gusts and lulls, and the latter can cause very rapid changes in supply.

Much of the growth of power generation in highly industrialized nations in recent years is in renewable supplies. But the penetration of those sources is fundamentally limited in a load-following regime (i.e., one in which power output is adjusted to demand).

Many smart grid proposals focus on increasing the capacity and sophistication of the transmission system to reduce constraints imposed by transmission in matching supply to demand. These include long-distance lines, in many cases using high-voltage DC, in order to access remote renewable supplies both for increased availability and to obtain geographic decorrelation. Within a grid, especially with distributed renewable resources, there may be sufficient supply to serve the load but inadequate capacity to route the power from points of generation to points of use. Better prediction, monitoring, and scheduling seek to prevent such bottlenecks. "Smart meters," which are currently being rolled out in many regions, provide 15-minute-interval readings, rather than monthly accounting. Their use enables more accurate prediction and more effective scheduling as well as introducing incentives, such as time-of-use pricing or critical peak pricing, to nudge the demand toward a more grid-friendly form. These efforts introduce a degree of observability into this complex system and thus open the way to decision making and action. As the IT in the grid evolves to embody monitoring, communication, embedded processing, and intelligence at various levels of the grid, it can provide a foundation for an interactive relationship between supply and demand that increases the penetration limit for renewable sources.

---

[19]For instance, Sandia National Laboratories just announced the development of a new family of liquid salt electrolytes that could lead to devices that could better incorporate renewable sources of energy on the grid. See, for instance, http://www.sciencecodex.com/sandia_national_laboratories_researchers_find_energy_storage_solutions_in_metils-86320.

Implementing a Supply-Following Model The dynamics and economics of grid operation can be fundamentally altered if the demand can be shaped to match available supply and supply-chain constraints, such as congestion or outages. This approach is referred to as supply following, in contrast to load following. Load following has developed over the past century with a great body of typically centralized, utility-side intelligence to permit consumers to use energy whenever and however they desire; however, supply following typically requires distributed, customer-side intelligence in order to manage energy demands while delivering desired services.

"Demand management" is typically taken to mean an explicit modulation of customer demand (for example, thermostat set-point adjustment, lighting adjustment, production curtailment) by the utility, according to prior arrangement, to atten the duration curve. Numerous such measures have been employed for peak shaving and shifting load into valleys, but adoption tends to be low. Prevalent utility-driven measures involve automated voluntary adjustment to thermostat set points, especially for cooling during hot summer days. Many industry-driven proposals emphasize smart appliances, including dishwashers, dryers, and ice makers that can defer operation until less costly times of use. Plug-in hybrid or fully electric vehicles are seen as presenting a prime opportunity for the programming of demand. While naïve charging could be potentially destabilizing to the grid or even cause local, aging distribution equipment to fail (for example, if multiple electric vehicles charge on the same block), well-timed charging could provide increased stability while relieving petroleum demand.[20]

Dynamic variable pricing (as opposed to set schedules) introduces financial incentives for end users to shift demand so that the overall demand is more easily met. Typically, residential demand is shifted into nights and weekends away from industrial demand. Often the schedules are complex and difficult for individual users to keep track of. Charging for power according to more sophisticated pricing schedules requires more sophisticated metering, along with usable and understandable

---

[20]There are also many proposals for utilizing electric vehicle batteries as grid storage, providing power back into the grid when demand is high. While certainly attractive in concept, such usage modes present pragmatic challenges. Batteries for vehicles are optimized to be extremely light, dense, and collision-resistant, with high power density. The number of recharge cycles of the battery that could occur before a costly battery replacement is a fundamental constraint. Utilizing this precious resource to improve the management of utility capital investment and potentially having driving range unexpectedly curtailed represents adoption challenges, whereas scheduling overnight charging attens the duration curve without such impediments. Stationary bulk energy storage need not obtain the very high level of energy density and power density demanded for the mobile case; it can potentially be designed instead for large recharge capacity and high turnaround efficiency.

controls. The modern advanced metering infrastructure rollout aims to achieve some of these possibilities.[21]

Peak pricing extends dynamic variable pricing to include aspects of a particular consumer's demand and to take into account how hard it is to meet that demand. In many regions, major industrial customers are on time-of-use, peak-based pricing schedules, in which the product of usage and peak demand determines the cost over a past interval. This creates an incentive for individual users to limit their peak as well as their overall demand.

Finally, an approach known as demand response typically involves load shedding in response to a critical peak-pricing notification event from the utility. Manual demand response has been used in many non-residential markets for many years. Automatic demand response couples an Internet notification event to a preprogrammed set of demand-mitigation responses and thus involves considerably greater IT. Significant information processing, modeling, and control issues need to be addressed to carry out demand-response issues in large commercial buildings. Significant human-computer interaction issues need to be addressed to realize fully the potential of this infrastructure, particularly for residential use. For example, if new electric water heaters are equipped with an automatic demand-response facility but the default setting is to ignore notifications, then without a good interface—and consumer education—the likely result is that the default setting will not be touched and there will be no benefit. However, if the default setting is to respond to a notification by waiting until late at night to turn the heater on, when electricity demand is lower, one can anticipate large numbers of consumers wondering why their hot water systems seem to run out of hot water at unexpected times.

Peak energy reduction is extremely important for reducing the capital investment in generation and transmission assets and in reducing risk in wholesale markets. Also, peak energy has the greatest GHG emissions per unit of electric power because it is generated by less-efficient plants with shorter ramp-up times. However, reducing peak energy has limited impact on reducing the overall energy demand or impact on climate, which are dominated by non-peak demand. Reducing overall demand and reducing the impact on climate require much broader efficiency and reduction measures. In some cases, low-carbon renewable generation—for example, summer solar production—aligns well with peak demand; in other cases, such as relying on the prevalence of nighttime wind, it does not.

---

[21]For more information on the deployment of advanced metering technologies, see National Energy Technology Laboratory, *Advanced Metering Infrastructure: NETL Modern Grid Strategy* (2008).

Accounting for Externalized Costs Another challenge to developing sustainable electric grids has to do with economic incentives and externalities. Various approaches have been articulated, including a carbon tax, fee and dividend models, and so-called cap-and-trade mechanisms (placing a cap on emissions but providing flexibility with mechanisms such as tradable permits). To truly represent the cost of externalities, any of these would require dramatically more precise accounting for environmental costs throughout the energy supply chain. In principle these offer a common metric around which optimization measures at all tiers can be integrated. Today, under a utility-centric approach, crude weightings of the energy blend appear to suffice. If a smarter grid were deployed, the task of accounting for all aspects of life-cycle costs would introduce tremendous IT challenges. The same would be true if, for instance, end users were to access information about the real-time mix in the blend to enable them to make more informed decisions about when to consume energy (assuming the presence of significant penetration of non-carbon supply).

Consideration of computer security should be integral to work on the smart grid. As just one example of the security risks, an attack that injected malicious code into smart electrical energy control systems in millions of homes might be used to manipulate demand and prices, or just to create chaos by turning on or shutting down large numbers of appliances unnecessarily. There are also legitimate potential privacy concerns with such control systems that will need to be addressed in ways that are both usable and technically sound.

Approaches to a More Sustainable Electric Grid

A forward-looking sustainable grid scenario presents a fundamentally more cooperative interaction between demand and supply and fundamentally greater transparency[22] throughout the energy supply chain, with the goal of achieving deep reduction in demand and deep penetration of renewables in the supply blend.[23]

---

[22]For this particular goal, transparency is required for technical reasons: that is, to support a more cooperative interaction between demand and supply. However, transparency is also essential for reasons of trust, accountability, and fairness, to avoid the potential for Enron-style market manipulations to be multiplied many times with the new grid technologies.

[23]Today, major industrial customers with substantial in-house generation are able to take advantage of the volatility of the wholesale market by running in-house generation when prices are high, tailoring its use to meet business goals, and shutting it down completely when prices "go negative"—which they do at times in the U.S. Midwest grid, in the European Union, and elsewhere—thereby getting paid to consume power.

The introduction of storage aims to decouple generation and consumption, enabling either to take place at times that are most effective. (But even with hypothetical vast storage capacities, inefficiencies in storing and generating electricity mean that complete decoupling will not be possible.) Equally important is the ability to coordinate loads, often through the utilization of non-electrical forms of energy storage such as thermal, but also often exploiting exibility in the actual task. In short, opportunities to improve the sustainability of the electric grid can be clustered as follows: (1) sculpt demand to match the supply of renewable power more closely; when abundant renewable power is available, use it for shiftable power needs (such as ice makers, hot water, dishwashers, electric car charging); (2) sculpt demand to smooth out spikes so that less high-speed dispatchable supply is needed; (3) reduce total demand by improved efficiency in transmission and use; and (4) apply instrumentation and modeling to measure carbon emissions as part of carbon pricing and capping policies.

Electricity is currently the most invisible of utility-supplied resources. To make progress on the opportunities available will require new forms of visibility, including visibility with respect to price and GHG emissions, consumption, and performance. IT can contribute in key ways: for example, pervasive instrumentation, monitoring, and analysis enable visibility into electric power consumption and resulting load performance.

Increased visibility is an important component of developing integrated home or building energy-management systems that can make wise decisions about how to shift energy use. These systems need exible user interfaces and sensing systems so that they can receive information about when it is appropriate to, say, delay electric car charging, heating up a building, and so on. For example, such systems could be integrated with users' calendars as well as connected to pricing and other generation-side signals (e.g., sun and wind forecasts). Research is needed on user interfaces, predictive models of user and appliance behavior, and perhaps auction and pricing mechanisms.

Integrated energy-management systems could also address spiking, by heating and cooling buildings more gradually or by making offsetting power-consumption choices. Making the effects of such choices visible to the user (e.g., by clarifying when the building will achieve the desired temperature, when it will be possible to turn on a particular machine, etc.) will be critical for user acceptance. Price mechanisms (e.g., charging more for sudden changes) could be explored as well.

IT research is needed for developing methods for sensing, modeling, and intelligent control of buildings. Most office buildings, for instance, exhibit chronic poor performance (some parts of the building are too hot, others too cold; some parts are poorly ventilated, others too breezy).

Improved modeling at design time and during operations has promise of reducing such problems and saving energy.[24] There is a significant role for IT to play in making opportunities for waste reduction manifest and in automating its reduction in buildings and also throughout the grid as a whole.

The Role of Information Technology and Computer Science in Achieving the Smart Electric Grid

Information and data management are essential to making progress toward a smarter, more sustainable electric grid, as discussed above. Computer science research and methodological approaches will be needed at all levels to address the broad systems challenges presented by the smart grid. Initial forays into both research and applications "wins" in this area include energy efficiency and smart mini-grid and distributed energy management,[25] energy-efficiency planning and building management,[26] and the integration of smart grids and smart electric vehicle planning and operation.[27] In many of these areas, savings of one-third to one-half in terms of overall energy consumption, with improved service and significant environmental gains, are possible. In many of these areas, a critical step is that of envisioning how the energy system could function if greater information and real-time data analysis were possible as embedded components of the system. This would require greater attention to integration of sensor technology with energy, transport, and building systems; to sensor data management; and to the role of distributed computing in processing far greater ows of information (and of forecasted performance and outcomes) than is typically the case today.

User interfaces are needed that make it straightforward for people to express preferences regarding aspects such as prices, comfort, timing, and "greenness" of their power mix. These preferences could be very complex and difficult to capture, requiring visualization techniques,

---

[24]National Research Council, *Achieving High-Performance Federal Facilities: Strategies and Approaches for Transformational Change*, Washington, D.C.: The National Academies Press (2011).

[25]C. Casillas and D.M. Kammen, The energy-poverty-climate nexus, *Science* 330:1181-1182 (2010).

[26]G. Crabtree L. Glicksman, D. Goldstein, D. Goldston, D. Greene, D.M. Kammen, M. Levine, M. Lubell, B. Richter, M. Savitz, and D. Sperling, *Energy Future: Think Efficiency—How America Can Look Within to Achieve Energy Security and Reduce Global Warming*, Report of the American Physical Society on the Potential for Energy Efficiency in a Low-Carbon Society, American Physical Society (2008).

[27]L. Schewel and D.M. Kammen, Smart Transportation: Synergizing Electrified Vehicles and Mobile Information Systems, *Environment: Science and Policy for Sustainability* 52(5): 24-35 (2010).

increased understanding of human behavior with regard to energy, pervasive interfaces, and so on. Information technologies may prove useful in encouraging energy consumers to shift their consumption patterns to off-peak hours, when consumption is generally more stable and comes from more sustainable sources of energy generation. In order to identify specific customers (with shiftable consumption patterns) to target with time-of-use rates, utilities could use more sophisticated predictive analysis through statistics. Information dispersal, social networking, and marketing through the Internet are other avenues that utilities are looking into, and with which computer science may be able to help. Additionally, with time-of-use rates, consumption data are increased significantly for each customer, and the complexity of tariffs and calculations with multiple tiers becomes an issue with which better analytical software could help.[28] Improved statistical models and database management would be invaluable additions to the capabilities of utilities all over the country.

Improved modeling and analytical tools would help with demand forecasting that takes into account the adaptive nature of the demands (e.g., to answer questions such as: How far will people be willing to time-shift demand this Thursday?). With the help of predictive analysis and weather data, utilities could use estimated capacity to improve their consumption forecasts, which would significantly improve their cost structure.

More generally, economic mechanism design tools for designing pricing within a controlled system will be needed. Sophisticated statistical models could help validate the models through hypothesis testing. A goal might be to bring factories with energy-production capacity into the supply chain to supplement peak-hour supply, assuming that the GHG output was not made worse. This goal would also require creating an economic situation that is agreeable to both the utilities and the owners of cogeneration plants. Calculating the cost and benefit from both the utilities' perspective and the cogenerators' perspective, optimizing the best rate scheme to encourage sell-back, and factoring in transmission losses and efficiency present a complicated and interesting optimization problem that could be greatly aided with the use of sophisticated decision analysis tools and statistical models.

Looking further ahead, if appropriate CS research is undertaken, the ability to mitigate the intermittency of renewables through computational approaches will be greatly enhanced. These resources could be made more dispatchable without the need for 1:1 matching of renewables and

---

[28]Shwetak Patel, University of Washington, described a technology to monitor in-home electricity consumption at the committee's Workshop on Innovation in Computing and Information Technology for Sustainability. See Appendix A for a summary of the workshop.

traditional sources or storage backup and optimized infrastructure investment. The success of the smart grid will, in part, be about the ability of the industry to shift from its current static operational, management, and planning models to a model that is increasingly dynamic—a scenario that CS research and IT are well poised to address. Whole new situational awareness tools are required to observe, monitor, and control the smart grid. The computational burden of doing this is significant, and the industry relies almost exclusively on vendors to supply solutions—vendors who typically do not invest a great deal in research and development. Switching from static balanced optimal power ow to dynamic transient analysis that can be solved in real time and at scale is not achievable today, but this ability will be a requirement for managing the future bidirectional, rapidly changing nature of the smart grid.

IT and CS approaches will have a fundamental role in aligning the temporal and spatial characteristics of resources and users and in reducing the need for the close alignment between supply and demand. Different methods and approaches will be needed for sustainable energy systems in small developing countries, for the micro-grid in developed countries, and for a continental-scale energy system. Major CS research is required to address these and related challenges.

Sustainable Food Systems

Agriculture in the United States and other parts of the world over the past century has been characterized by a dramatic increase in productivity, resulting in relatively affordable and available food. Some of the driving factors for this increase include the relatively inexpensive availability of fossil fuels and abundant fertilizer and water; concentration and specialization in farm production, including the increased use of automation and robotics in meat processing; the increased mechanization of farming and the availability of new technologies; advances in plant breeding; government programs and subsidies; and the expansion and commercialization of markets. These developments have allowed for food to be produced at unprecedented volumes and have supported significant population growth.[29] The increase in output from agriculture, unlike that in many other industries, has not been associated with a similar increase in inputs. For example, the acreage of cropland used in 2005 was comparable to

---

[29]Indeed, a recent issue of *The Economist* carried a "debate" about whether computing was the most significant technological advance of the 20th century, and the "anti" side, articulated by Vaclev Smil, argued that the transformation of agriculture enabled by the production of sufficient nitrogen for fertilizer with the Haber-Bosch process of nitrogen fixation was much more significant. See http://www.economist.com/debate/days/view/598.

the acreage used in 1910. Thus, there has been a tremendous increase in productivity in U.S. agriculture, with more food being produced with significantly less capital, land, labor, and materials.[30]

However, current agricultural practices pose challenges to the sustainability of the food system, as well as to the broader social, economic, and environmental systems within which they are embedded. Much of the focus of agriculture has been on maximizing production to satisfy human food, feed, and fiber needs while secondarily considering environmental and societal impacts. There is a growing concern regarding the negative consequences of current trends in agricultural productivity and a concern that these trends cannot continue indefinitely. Increases in agricultural productivity have spawned hypoxia in coastal and inland waters around the world because of increased concentrations of nitrogen and phosphorus, altering the planet's biogeochemistry. A sustainable food system will be key to ensuring that the world's population receives necessary nutrition without contributing additional damage to the environment and society. As with the electric grid, the opportunities for IT seem most salient in the systems issues in sustainable agriculture.

The recent report of the National Research Council on sustainable agriculture defines a "sustainable agriculture system" as one that (1) satisfies human food, feed, and fiber needs and contributes to biofuel; (2) enhances environmental quality and the resource base; (3) sustains the economic viability of agriculture; and (4) enhances the quality of life of farmers, farmworkers, and society as a whole.[31] The first point naturally requires both sufficient food production and a population sized appropriately for the food that can be produced. The American Public Health Association, in its policy statement on sustainable food systems, builds on these ideas, defining a sustainable food system as one that "provides healthy food to meet current food needs while maintaining healthy ecosystems that can also provide food for generations to come with minimal negative impact to the environment."[32]

A sustainable food system will need to address simultaneously all four of the objectives listed above rather than optimizing over any individual dimension. A 2012 NRC report of two workshops provides a broad exploration of food security, agriculture, and related sustainability chal-

---

[30]National Research Council, *Toward Sustainable Agricultural Systems in the 21st Century*, Washington, D.C.: The National Academies Press (2010).
[31]National Research Council, *Toward Sustainable Agricultural Systems in the 21st Century*, Washington, D.C.: The National Academies Press (2010).
[32]American Public Health Association, "Toward a Healthy, Sustainable Food System" (2007). Available at http://www.apha.org/advocacy/policy/policysearch/default.htm?id=1361.

lenges. It notes that "neither the modern food systems nor the traditional systems assure long term food security for all" and examines availability, access, and utilization as well as barriers to expanding production (without damaging future capacity) and policy, technology, and governance interventions that could help.[33]

This section brie y explores the challenges facing the creation of a sustainable food system that promotes public health and identifies potential areas in which IT and research in computer science can have an impact.

Challenges to Developing a Sustainable Food System

A few of the many challenges to creating a sustainable global food system are highlighted below. They include increasing demand, environmental impacts, and public health impacts.

Increasing Demand The U.S. population increased by approximately 9 percent from 2000 to 2009, and total consumption of food has increased in parallel. In addition to the increase in the total amount of food consumed, the composition of the nation's diet has shifted toward an increased consumption of meat beyond levels recommended by federal guidelines. Since 1960, there has been an increase in the use of grains for livestock feed, and so a shift toward meat consumption produces a greater strain on the agricultural system.[34] Rising incomes in emerging markets such as Mexico and China have produced greater demands on U.S. agricultural exports, and such demand is likely to increase in the future. The emerging biofuels and bioenergy fields have also placed further demands on agriculture to provide materials for alternative energy production. In 2007 and 2008, 23 percent of the U.S. corn harvest and 17 percent of the soybean harvest were used to produce ethanol and biodiesel. The various demands on agriculture today increasingly strain the natural resources of land and water that are required to satisfy global food needs.

Environmental Impacts Agriculture is a contributor to greenhouse gas (primarily methane and nitrous oxide) emissions through various soil-management activities and livestock operations. Through biomass burning and windblown dust, farms also serve as sources of air pollutants, such as particulate matter. Conventional industrial agriculture applies

---

[33]National Research Council, *A Sustainability Challenge: Food Security for All: Report of Two Workshops,* Washington D.C.: The National Academies Press (2012), p. 2.
[34]Food and Agriculture Organization of the United Nation, "Livestock's Long Shadow" (2006). Available at http://www.fao.org/docrep/010/a0701e/a0701e00.htm.

large amounts of nitrogen-based fertilizers in order to replenish nutrients in the soil and substantial quantities of herbicides and pesticides to control both plant and insect pests. The increased use of such fertilizers and pesticides leads to runoff during floods or heavy rains, which pollutes rivers, streams, and bays. Tilling of the soil contributes to land degradation, and farming in dry regions consumes water resources for irrigation. It is estimated that in the United States, 80 percent of available potable water is used for agricultural irrigation, and overdrafting of underground aquifers (when the rate of extraction exceeds the rate of natural recharge) threatens agricultural activity in a large swath of the U.S. Midwest.[35]

Public Health Impacts In addition to the environmental impacts of modern agriculture, there are public health impacts from current agricultural practices. Air and water pollution from farms damages not just the environment but also the health of the individuals living and working in or near the damaged environment. Factory farming of food animals has also increased the risk of foodborne pathogens, in part due to the close quarters of animals kept in confined animal feeding operations (CAFOs). The interplay between supply and demand of highly processed foods has health implications as well, even as debate continues about the specific mechanisms and contributors to diseases such as diabetes and heart disease.

Approaches to Developing Sustainable Food Systems

Creating a more sustainable global food system will not be easy. This section outlines several approaches, none of which is sufficient alone, although each contributes to increased sustainability and could benefit from the contributions of computer science and IT. The approaches include taking a systems view; developing methods for measuring the costs, benefits, and impacts of different agricultural systems; the use of precision agriculture; information for informed consumption; and the development of social networks for local food sourcing.

Taking a Systems View Overall, there is a need to take a systems view of agriculture (much like taking a systems view in other areas of sustainability, such as the smart grid, described previously) in order to understand and analyze the total impact of agriculture on the environment, economy, and society. A systems perspective is relevant at all points in the sys-

---

[35]National Research Council, *Toward Sustainable Agricultural Systems in the 21st Century*, Washington, D.C.: The National Academies Press (2010). Also see http://ga.water.usgs.gov/edu/wuir.html.

tem. At the farm itself, individual farms can combine crop and livestock production so as to reduce the need for synthetic fertilizers. Traditional agriculture has focused on controlling the farm ecosystem by simplifying it (e.g., through monoculture) and applying external inputs (e.g., water, fertilizer, pesticides, and herbicides). The systems view seeks to reduce or eliminate those external inputs (and their associated carbon and pollution emissions) by, for instance, designing and managing a more complex ecosystem involving a larger variety of species.[36] A systems view can provide guidance on how to develop ways to make the system as a whole more sustainable. For instance, rather than viewing a farm (or farms) in isolation and having inputs and outputs, one could view the entire cycle of food production and consumption as providing natural resources for growing food that is consumed by people. This cycle includes land, water, and other farm inputs, crops, transportation, processing, retailing, consumption, and recycling or waste. At each stage, there are effects on the environment and society; thus it is important to consider the connections between farms, the ecosystem, and communities (local, regional, and global). An important role for IT is to enable farmers to manage these more complex systems through mechanisms such as sensing, predictive modeling, and precision machinery.

Methodology for Measuring Costs, Benefits, and Impacts There is a substantial need for the development of methods and tools to measure the total costs, benefits, and impacts of different agricultural systems. For example, comparative studies of GHG emissions from different field-management practices for animal wastes would allow for better quantification of the environmental impacts of agricultural systems and, just as with the smart grid scenario, allow for prices to re ect costs and value better. In general, evaluating different farming systems will require assessing how each system balances productivity and efficiency with environmental and societal impacts and will require analyzing the behavior of complex high-dimensional and highly interactive systems. In addition to the technical challenges of developing such measures, there are also significant challenges in helping them to be seen as accurate and legitimate by both producers and consumers. Novel visualization techniques, explanation facilities, interactive simulations, and other techniques may help here.

---

[36]The control of pests provides an example of moving from a traditional view to a systems view. The traditional way of controlling pests is to apply pesticides, which requires little knowledge of the pests. A more sustainable approach may be to use benign control measures, which require an understanding of the pest's life cycle and its interaction with other parts of the farm ecosystem.

Precision Agriculture The use of information and computing technology in agriculture has greatly increased in the past 50 years. It has allowed farmers to assess variation within fields and to generally maintain or increase yields while reducing inputs (particularly water, nutrient, and pesticide application). Technologies used here include the Global Positioning System, real-time kinematics, and geographic information systems, especially satellites. IT already plays a substantial role in this area and will continue to play a critical role in the future. There is also a connection with methodologies for measuring costs and benefits: if the cost of water for agricultural use re ects its true cost, there may be much more incentive to use precision agriculture to reduce the consumption of water.

Information for Informed Consumption Increasing the information available to individuals regarding the nature of the food that they buy and how it was produced can assist them in making sustainable choices about food. Already there is an emerging market for foods that have been produced in a sustainable manner.[37] An important method by which such information is currently conveyed is through the development of standards, certifications, or other eco-label programs. Each of these programs outlines a set of criteria for food producers and distributors in an effort to address various environmental, sustainability, or health goals.[38] Perhaps the most well-established food standard in the United States is the organic agriculture certification, which focuses primarily on health and environmental goals and does not address the broader goals of sustainable agriculture. Food-labeling requirements in the United States provide some information, for example, on the country of origin of meats and fruits, but general information about sustainability and food transport (which has implications for fossil fuel usage) is not available. Current standards and certifications are typically communicated using logos or other print labeling on food packaging. However, potential exists for providing much richer information regarding sustainability and information to help consumers sort through the proliferation of eco-labels in the market. The wider adoption of smartphones may allow for easier dissemination of this information, as users could search for sustainability information at the point of purchase. One example of empowering individuals with information is the Monterey Bay Aquarium's Seafood Watch guide[39] that

---

[37]National Research Council, *Toward Sustainable Agricultural Systems in the 21st Century*, Washington, D.C: The National Academies Press (2010). Chapter 6.
[38]Ecolabel Index. Available at http://www.ecolabelindex.com/.
[39]The Monterey Bay Aquarium's Seafood Watch guide is available at http://www.montereybayaquarium.org/cr/seafoodwatch.aspx. The guide provides a list

provides detailed and up-to-date information about what types of seafood are caught or farmed in a sustainable manner. In addition to a web site, the guide also is available as a smartphone application so that consumers can have portable access to its vast database.

Social Networks for Local Food Sourcing IT could be used to increase networking among individuals and organizations, encouraging locally and regionally sourced food consumption. Community-supported agriculture (CSA) already benefits from the organizing power of online networks to distribute relevant information, create markets for local farm producers, make it easier to place orders, and help connect consumers with local food. Generally, IT could be used to help make a more effective market for local foods.[40] Beyond efficiency, there is little argument that humans have emotional connections to food; techniques to strengthen the farmer/consumer connection could also be valuable. IT could also be useful for gathering information on regional surpluses or deficits, allowing fresh foods to be allocated to areas where they are most needed and diminishing reliance on processed foods with longer shelf lives.

The Role of Information Technology and Computer Science in Achieving a Sustainable Food System

As with the smart electric grid, information and data management are essential to making progress toward a smarter, more sustainable, global food system. Computer science research and methodological approaches will be needed at all levels to address the broad systems challenges—encompassing the environment and ecosystems, social and economic factors, and personal and organizational behaviors—affecting food production, distribution, and consumption. Three critical areas are described brie y below: information integration; education and reform; and systems modeling, prediction, and optimization.

Information Integration Information integration can help individuals and organizations on both the demand and the supply side of the food system

---

of sustainable choices and the least sustainable choice of fish to consume. Legal Sea Foods has questioned the value of the guide. See http://www.nrn.com/article/legal-sea-foods-defies-aquarium%E2%80%99s-watch-list.

[40]As one example of an effort in this area, see http://www.urbaninformatics.net/projects/food/ regarding a project exploring "ubiquitous technology for sustainable food culture in the city." Another example is LocallyGrown.net, which seeks to provide an online infrastructure and organizational capacity for local farmers' markets and CSAs, particularly small-scale growers with few or no employees.

make sustainable choices regarding the production and consumption of food. Providing consumers with information about the sustainability of food production, in addition to other aspects of sustainable food systems such as health and environmental impacts, will require the sharing and integration of information across producer and consumer platforms. Developing and optimizing the infrastructure and architecture for such information integration will be an important contribution of IT. More generally, areas in which IT could be of substantial help include the creation of databases of information and the maintenance of the currency of that information as well as connecting farmers and consumers through social networks and the Internet. The development of analytical software for optimizing sustainable food purchasing choices for both consumers and large-scale purchasers (such as supermarkets) is another rich area of IT contributions.

Education and Reform Tools are needed to help both consumers and policy makers understand the trade-offs posed by the global food system and to navigate those trade-offs toward increased sustainability. The role of IT here is not just in providing information on availability and techniques, but also in allowing access to communities of individuals with similar interests. There are numerous opportunities to effect change through demand-side modification of food consumption.[41] Efforts to encourage the preparation and even the growing of food at home could have a significant impact on overall distribution needs. Increasing the availability of fresh, healthful foods in certain communities (e.g. low-income communities) would also help. Additional challenges exist in predicting the information that purchasers and individuals will need, displaying information that will encourage more sustainable consumption habits, educating consumers about sustainable choices without overwhelming them, and so on.

Systems Modeling, Prediction, and Optimization Improving the efficiency of the food system in general will require modeling a complex and interactive system and methods for predicting food shortages and surpluses in order to help ensure that food is available in different regions at

---

[41] Andrea Grimes, Martin Bednar, Jay David Bolter, and Rebecca E. Grinter, EatWell: Sharing nutrition-related memories in a low-income community, *Proceedings of the 2008 ACM Conference on Computer Supported Cooperative Work* (2008); Andrea Grimes and Richard Harper, Celebratory technology: New directions for food research, *Proceedings of the Twenty-Sixth Annual SIGCHI Conference on Human Factors in Computing Systems* (2008); and T. Aleahmad, A. Balakrishnan, J. Wong, S. Fussel, and S. Kiesler, Fishing for sustainability: The effects of indirect and direct persuasion, *Extended Abstracts from Conference on Human Factors in Computing Systems* (2008).

reasonable costs. In addition, the transportation of food to various markets could be optimized according to sustainability cost functions if a comprehensive model of the food system were available. Given a model of the food system, one could also assess the costs and benefits of various agricultural and farming strategies, the design of food sheds, and distribution systems.

Sustainable and Resilient Infrastructures

The resilience of the nation's societal and physical infrastructures poses deep and crosscutting sustainability challenges, especially when one takes a broad view of sustainability that encompasses economic and social issues. For example, although transportation is a major source of GHG emissions and urban sprawl consumes open space and farmland, competing incentives in the realm of societal sustainability include the need for workers to commute to jobs, for people to have access to whole foods, and for available space that allow businesses to change and adapt over time. Contributing to the challenges of resilience of societal and physical infrastructures is the increasing risk of natural and human-made disasters. Sustainability concerns related to climate change, resource consumption, and land use are closely linked to natural and human-made disasters.[42] There will inevitably be more disasters, and enhancing society's resilience and ability to cope with them will contribute to sustainability. Even apart from climate and resource consumption, the sheer magnitude of the world's population means that crises, when they happen, will be at larger scale. This section examines the sustainability challenges around planning and modeling infrastructure and anticipating and responding to increasing disasters and the ways in which information technology can assist with developing sustainable and resilient infrastructures. The section focuses on cities as centers of large human populations, but many of the issues discussed apply generally.

Challenges to Developing More Sustainable and Resilient Infrastructures

Cities are highly complex, evolving systems, involving the interaction of numerous people and processes, as well as natural and built infrastructure, legal and regulatory frameworks, and much else. The diversity of use within the systems adds another level of complexity. Each building's use and design are unique within a particular city; each city's infrastructure has distinctive characteristics. The heterogeneity of structures within any

---

[42]National Research Council, *Adapting to the Impacts of Climate Change*, Washington, D.C.: The National Academies Press (2010).

given city poses challenges—and because cities are often quite different from one another, the extrapolation of lessons learned is also challenging.

Just as cities are increasingly complicated, the challenges of coping with disasters are compounded by their heterogeneity. There are acute natural disasters (such as hurricanes, earthquakes, and floods), acute engineering and other human-made disasters (such as the 2010 Deepwater Horizon Gulf oil spill), as well as "slow" or chronic disasters (such as droughts, refugee crises, and rising sea levels). In addition, whether acute or chronic, there are the ongoing processes of cleanup and recovery from disasters. Many situations are best described as combinations of natural and human-made disasters with both acute and chronic time frames.[43]

The problems associated with the resilience of societal and physical infrastructures have complicated time lines. For instance, urban, suburban, and rural areas are developed over long periods of time and are almost constantly being shifted into new uses. These long time lines create legacy systems that may not be compatible with newer systems or that could be costly to update. Planning becomes increasingly complicated as new infrastructure, often costly and time-consuming to implement, must anticipate the future needs of a particular area.

Similarly, the time needed and the ability to prepare an area for potential emergencies vary and depend not just on characteristics of the area, but also on the anticipated types of disasters and crises. Some disasters, like hurricanes, come with at least some advance warning, and others, like earthquakes, strike at unpredictable times. Some events cause intense damage only in limited areas, while others affect enormous geographical regions. An additional challenge is that the frequency of disastrous events is such that recovery after one event (itself a major sustainability challenge) may well not be complete before the next major disaster strikes—either in the same region, as happened with Hurricane Katrina and the Deepwater Horizon oil spill, or different regions competing for resources and attention, as in the earthquake in Haiti in 2010 that was followed by severe flooding in Pakistan.

The Role of Information Technology in Developing Sustainable and Resilient Infrastructures

Information and communications technologies offer a range of methodologies, approaches, applications, and tools that will be integral to the

---

[43] Author Bruce Sterling coined the term "Wexelblat Disaster" to refer to disasters caused by the interaction of natural disasters and failures of human-engineered technology. The 2011 earthquake and tsunami that destroyed a nuclear power plant in Japan leading to core meltdowns is an example.

development of sustainable and resilient infrastructure and to coping with disasters when they occur. Several such technologies are highlighted below.

Modeling and Simulation Urban regions can be modeled with varying degrees of spatial detail and behavioral realism. For a highly disaggregate, behaviorally realistic model,[44] the process of modeling a new region is time-consuming, often requiring person-years of effort. A major factor is difficulties in collecting and readying the needed data. Further, problems of missing data—common in U.S. metropolitan regions and even more so in developing countries—make the task much more challenging. Modeling the development of cities over periods of 20 or more years, under different alternatives, can provide important information to inform public deliberation and debate about alternate plans and possible futures. Transportation modeling, and more comprehensively integrated modeling of urban land use, transportation, and environmental impacts, have a substantial history and are in operational use in many regions. Nevertheless, there are major limitations in current knowledge, and new research is needed to address the coming challenges adequately. In addition to the scientific challenges of the modeling itself, it is important to consider how the modeling work fits into the larger political and organizational process of making major decisions (often a contentious process), and to shape the technology to respond to these contextual challenges.

Turning from simulations of long-term development to immediate support for coping with disasters: during a disaster copious amounts of information can be collected; however, more does not always mean better or more helpful information.[45] Sorting out how to manage and use IT capabilities at hand most effectively and, perhaps even more importantly, the vast amounts of data that can be made available by those capabilities, is a non-trivial exercise.[46,47]

---

[44]For example, UrbanSim (http://www.urbansim.org), currently the most widely employed land use model in the United States.

[45]Bruce Lindsay, *Social Media and Disasters: Current Uses, Future Options, and Policy Considerations*, Congressional Research Service (2010). Available at http://www.fas.org/sgp/crs/homesec/R41987.pdf.

[46]See "Disaster Relief 2.0: The Future of Information Sharing in Humanitarian Emergencies," available at http://www.unocha.org/top-stories/all-stories/disaster-relief-20-future-information-sharing-humanitarian-emergencies, for an early assessment of crowdsourcing information and data ows in a humanitarian crisis. In this case the Haiti earthquake of 2010 was a primary example.

[47]Dan Reed, vice president of Microsoft Research, discussed some of the computational challenges posed by the 2010 Gulf oil spill, noting that the disaster stemmed from a "complex multidisciplinary system with emergent behaviors across a wide range of temporal and spatial

In addition to modeling the effects of current disasters, IT offers opportunities for in-depth simulation of potential disasters and for individuals to exercise and manage a given organization's response to a crisis to hone and refine their skills and approach.

Communication IT provides the communications capabilities before, during, and after a crisis for coordinating activities and for delivering alerts and warnings to affected populations. IT provides critical capabilities for the other phases of crisis response as well, such as modeling and simulation to predict likely consequences or to contribute to the understanding of the effectiveness of particular mitigation measures. As discussed in a 2007 NRC report, IT provides capabilities that can help people make better sense of information, grasp the dynamic realities of a disaster more clearly, and help them formulate better decisions more quickly. IT provides the tools to capture knowledge and share it with disaster-management professionals and the public. IT can help keep better track of the myriad details involved in all phases of disaster management.[48]

The Role of Information Technology and Computer Science Research in Developing Sustainable Infrastructure and Fostering Resilience

Advances will be needed in IT and computer science research and methodological approaches to enable better simulations and better understanding of the uncertainties associated with achieving more sustainable development that is also more resilient in the face of disaster. Advances are also needed in the areas of encouraging citizen participation, developing indicators of resilience and future outcomes, and improving IT infrastructures themselves.

Performance Running a simulation for a high-end, behaviorally realistic model for a major metropolitan region is a slow process, currently often requiring days, even on today's fast computers. Similarly, the process of constructing a new scenario (i.e., a package of infrastructure improvements, zoning changes, tax incentives, and perhaps such things as tolling

---

scales." He described some of the challenges in modeling such a system: "we lack the software engineering and programming methodologies to assemble, test and verify an integrated solution . . . the computational demands of an integrated, fully multidisciplinary, parametric simulation study of the oil spill and its effects would make accurate climate modeling seem like child's play on an abacus by comparison." Dan Reed, Lessons from the Gulf of Mexico (2010), available at http://www.hpcdan.org/reeds_ruminations/2010/08/lessons-from-the-gulf-of-mexico.html.

[48]National Research Council, *Improving Disaster Management: The Role of IT in Mitigation, Preparedness, Response, and Recovery*, Washington, D.C.: The National Academies Press (2007).

or congestion pricing) can require months of work by experts in transportation, land use modeling, and other disciplines. Creating far more efficient algorithms for modeling systems that are increasingly complex presents an important challenge. One natural approach is parallelizing the algorithms. In a number of cases this is quite feasible: for example, when modeling residential location choice (where will people decide to live, given characteristics both of the household and the possible dwellings), one can use massive parallelism, with each household making its decisions independently. One must then undo some assignments if two households attempt to move into the same place simultaneously (perhaps mirroring what happens in real life with several people all trying to rent or buy the same dwelling). However, new or improved algorithms are likely a richer source of performance gain, which will be important because many of the applications envisioned require huge performance increases (for example, using a simulation in real time in a meeting, or running a simulation many, many times to compute information about uncertainty). The precomputing of key scenarios and interpolating among the results (when the changes are smooth rather than abrupt), rather than computing the results from each scenario from scratch, should also be investigated. In terms of algorithms, one class of new algorithms that should be investigated is multiscale models, in which the simulation is first run at a relatively coarse grain (e.g., a zonal level), and the results from this are fed to further simulation runs within each zone, and so forth. (See Chapter 2 for more on modeling.) In this case the reason for using a multiscale model is performance. Heterogeneous models are also relevant for urban simulation—for example, coupling UrbanSim (a regional-scale model) with statewide freight mobility models. This could be further optimized by simulating only within zones that have changed significantly from the prior simulation period or that are of particular policy interest, and otherwise remaining at the coarser level.

Managing Uncertainties Urban modeling is rich with uncertainties on many levels, including future population, global economic conditions, the price of energy, the impact of climate change, and many others. There have been some successes in propagating uncertainty through the modeling process and capturing it in the indicators that the system produced,[49] but much more needs to be done in terms of both statistical techniques and effective presentation of the results.

---

[49]Hana Ševcíková, A. Raftery, and P. Waddell, Assessing uncertainty in urban simulations using Bayesian melding, *Transportation Research Part B: Methodology* 4:652-659 (2007).

Citizen Participation To date, citizen science (or citizen information gathering) is being used for such activities as open mapping projects, but much of this type of activity has not been integrated with modeling work. Harnessing the energies and interests of citizen scientists has strong potential, both as a source of additional data and as an avenue for public participation and the legitimation of the modeling activity. Leveraging existing technology (such as mobile applications, cloud services, mapping and location services, microcommunications platforms, social media, and so on) offers numerous opportunities to improve approaches to emergency and disaster management.[50]

Some organizations are experimenting with gathering situational awareness from citizens, and in particular citizen use of social media.[51] At the same time, there are significant challenges with regard to data quality, coverage, and institutional acceptance, among other things. Technical approaches here may include reputation systems that let staff at institutions build up confidence in particular observers, and ways to correlate data from multiple observers and to detect outliers.

During disasters, more attention should be paid to the information and resources held by the public because members of the public collectively have a richer view of a disaster situation, may possess increasingly sophisticated technology to capture and communicate information, and are an important source of volunteers, supplies, and equipment. Again, the information provided by the public will not always be correct; further, making full use of it may require considerable changes to existing practices. It is likely that the development of new, automated, and mixed-initiative techniques to manage and process the potential ood of information will be needed. Another important factor is how to engage the entire population, given the existence of groups with cultural and language differences and other special needs.

Indicators of Future Outcomes Simulations already produce indicators of such outcomes as GHG emissions, consumption of open space, and comparative measures of compact versus low-density development, all for multiple years and under different scenarios. However, as discussed above, it is also necessary to anticipate disruptions and potentially even disasters, due to climate change, mass movement of refugees, and other

---

[50]National Research Council, *Public Response to Alerts and Warnings on Mobile Devices: Summary of a Workshop on Current Knowledge and Research Gaps*, Washington, D.C.: The National Academies Press (2011).

[51]Sarah Vieweg, Amanda Hughes, Kate Starbird, and Leysia Palen, Microblogging during two natural hazards events: What Twitter may contribute to situational awareness, *Proceedings of the 2010 ACM Conference on Computer Human Interaction*, pp. 1079-1088.

factors. A research challenge is to develop indicators of community resilience in the face of such events.[52] These might include the percentage of electrical energy generated locally (or that could be generated locally if need be), the redundancy of the transportation system and the food supply chain and their ability to cope with a sharp increase in fuel prices or even rationing, the ability to cope with sea-level rise (if relevant), the ability to walk to the most significant destinations if need be, the availability of food produced nearby, and so forth. These indicators need to be accepted by decision makers and the community to be useful in the political process. More abstract and much more difficult, if not impossible, to incorporate into a predictive model (but nevertheless important) are the civic capital and connectedness of the community.

IT Infrastructure Improvements Large disasters upset physical infrastructure, such as the electric grid, transportation, and health care—as well as IT systems. IT infrastructures themselves need to be more resilient; IT can also improve the survivability and can speed the recovery of other infrastructure by providing better information about the status of systems and advance warning of impending failures. Finally, IT can facilitate the continuity of disrupted societal functions by providing new tools for reconnecting families, friends, organizations, and communities.

CONCLUSION

IT and computer science could have a major impact in a wide diversity of sustainability challenges. The examples above illustrate some of the efforts that are needed. Individual problems are highly multidimensional, requiring innovation in different areas of computing as well as deep domain knowledge.

> **FINDING: Although sustainability covers a broad range of domains, most sustainability issues share challenges of architecture, scale, heterogeneity, interconnection, optimization, and human interaction with systems, each of which is also a problem central to CS research.**

The next chapter explores more specifically the potential for computing and IT research and innovation to help address these challenges.

---

[52] An example of this is the Climate Change Habitability Index. For a description, see Yue Pan, Chit Meng Cheong, and Eli Blevis, The Climate Change Habitability Index, *Interactions* 17(6):29-33 (2010).

2

Elements of a Computer Science Research Agenda for Sustainability

The discussion of sustainability challenges in Chapter 1 shows that there are numerous opportunities for information technology (IT) to have an impact on these global challenges. A chief goal of computer science (CS) in sustainability can be viewed as that of informing, supporting, facilitating, and sometimes automating decision making—decision making which leads to actions that will have significant impacts on achieving sustainability objectives. The committee uses the term "decision making" in a broad sense—encompassing individual behaviors, organizational activities, and policy making. Informed decisions and their associated actions are at the root of all of these activities.

A key to enabling information-driven decision making is to establish models and feed them with measurement data. Various algorithmic approaches, such as optimization or triggers, can be used to support and automate decisions and to drive action. Sensing—that is, taking and collecting measurements—is a core component of this approach. In many cases, models are established on the basis of previous work in the various natural sciences. However, in many cases such models have yet to be developed, or existing models are insufficient to support decision making and need to be refined. To discover models, multiple dimensions of data need to be analyzed, either for the testing of a hypothesis or the establishing of a hypothesis through the identification of relationships among various dimensions of measured data. Data-analysis and data-mining tools—some existing and some to be developed—can assist with this task.

Once a model is established, "what-if" scenarios can be simulated, evaluated, and used as input for decision making. Modeling and simulation tools vary widely, from spreadsheets to highly sophisticated modeling environments. When a model reaches a certain maturity and trust level, algorithms, such as optimizations or triggers, can be deployed to automate the decision making if automation is appropriate (for example, in terms of actuation). Alternatively, information can be distilled and presented in visual, interactive, or otherwise usable ways so that other agents—individuals, organizations and businesses, and policy makers and governments—can deliberate, coordinate, and ultimately make appropriate, better-optimized choices and, ultimately, actions.

All of the steps described above must be done in an iterative fashion. Given that most sustainability challenges involve complex, interacting systems of systems undergoing constant change, all aspects of sensing, modeling, and action need to be refined, revised, or transformed as new information and deeper understandings are gained. A strong approach is to deploy technology in the field using reasonably well understood techniques to explore the space and to map where there are gaps needing work. Existing data and models then help provide context for developing qualitatively new techniques and technologies for even better solutions.

> **FINDING: Enabling and informing actions and decision making by both machines and humans are key components of what CS and IT contribute to sustainability objectives, and they demand advances in a number of topics related to human-computer interaction. Such topics include the presentation of complex and uncertain information in useful, actionable ways; the improvement of interfaces for interacting with very complex systems; and ongoing advances in understanding how such systems interact with individuals, organizations, and existing practices.**

Many aspects of computer science and computer science research are relevant to these challenges. In this chapter, the committee describes four broad research areas, listed below, that can be viewed as organizing themes for research programs and that have the potential for significant positive impact on sustainability. The list is not prioritized. Efforts in all of the areas will be needed, often in tandem.

- Measurement and instrumentation;
- Information-intensive systems;
- Analysis, modeling, simulation, and optimization; and
- Human-centered systems.

For each area, examples of research problems focused on sustainability opportunities are given. The discussions do not provide a comprehensive list of problems to be solved, but do provide exemplars of the type of work that both advances computer science and has the potential to advance sustainability objectives significantly. In examining opportunities for research in CS and sustainability, questions that one should attempt to answer include these: What is the potential impact for sustainability? What is the level of CS innovation needed to make meaningful progress?

As discussed in Chapter 1, complete solutions to global sustainability challenges will require deep economic, political, and cultural changes. With regard to those changes, the potential role for CS and IT research discussed in this chapter is often indirect, but it is still important. For example, CS research could focus on innovative ways for citizens to deliberate over and to engage with government and with one another about these issues, with the deliberations closely grounded in data and scientific theory. For some critical sustainability challenges, such as the anticipated effects of global population growth, the potential CS research contribution is almost entirely of this indirect character. For instance, there is potential for using the results of modeling and visualization research toward the aim of improved education and better understanding of population and related issues. In addition, advances in IT in the areas of remote sensing, network connectivity services, adaptive architectures, and approaches for enhanced health diagnosis and care delivery—especially in rural areas— also have a bearing on population concerns. Other contributions from CS and IT research toward meeting such challenges could be aimed at developing tools to support thoughtful deliberation, with particular emphasis on encompassing widely differing views and perspectives.

The research areas described in this chapter correspond well with the broader topics of measurement, data mining, modeling, control, and human-computer interaction, which are, of course, well-established research areas in computer science. This overlap with established research areas has positive implications—in particular, the fact that research communities are already established making it unnecessary to develop entirely new areas of investigation. At the same time, the committee believes that there is real opportunity in these areas for significant impacts on global sustainability challenges. Finding a way to achieve such impacts effectively may require new approaches to these problems and almost certainly new ways of conducting research.

In terms of a broad research program, an important question is how to structure a portfolio that spans a range of fundamental questions, pilot efforts, and deployed technologies while maintaining focus on sustainability objectives. For any given research area in the sustainability space, efforts can have an impact in a spectrum of ways. First, one can explore the immediate applicability of known techniques: What things do we know how to do already with computational techniques and tools, and how can we immediately apply them to a given sustainability challenge? Second, one can seek opportunities to apply known techniques in innovative ways: Where are the opportunities in which the straightforward application of a known technique will not work but where it seems promising to transform or translate a known technique into the domain of a particular sustainability challenge? This process tends to transform the techniques themselves into new forms. Finally, one can search for the areas in which innovation and the development of fundamentally new computer science techniques, tools, and methodologies are needed to meet sustainability challenges. While endorsing approaches across this spectrum, the committee urges emphasis on solutions that have the potential for significant impacts and urges the avoidance of simply developing or improving technology for its own sake.

The advancing of sustainability objectives is central to the research agenda outlined in this report. As in any solution-oriented research space, there is a tension between solving a substantive domain problem, perhaps creating tools, techniques, and methods that are particularly germane to the domain, and tackling generalized problems, perhaps motivated by the domain, for which solutions advance the broader field. (Chapter 3 discusses this challenge in more detail and provides the committee's recommendations on structuring research programs and developing research communities in ways that constructively address these issues.) When focusing on the challenges presented in a particular domain, it is often essential that the details are right in order for the work to have meaningful impact. For the work to have broader impact, it must be possible to transcend the details of a particular problem and setting. Much of the power in computer science derives from the development of appropriate abstractions that capture essential characteristics, hide unnecessary detail, and permit solutions to subproblems to be composed into solutions to larger problems. A focus on getting the abstraction right for large impact, appropriability, and generalizability is important. Simultaneously, it is important to characterize aspects of the solution that are not generalizable.

> **FINDING: Although current technologies can and should be put to immediate use, CS research and IT innovation will be critical to meeting sustainability challenges. Effectively realizing the potential of CS to address sustainability challenges will require sustained and appropriately structured and tailored investments in CS research.**

PRINCIPLE: A CS research agenda to address sustainability should incorporate sustained effort in measurement and instrumentation; information-intensive systems; analysis, modeling, simulation, and optimization; and human-centered systems.

MEASUREMENT AND INSTRUMENTATION

Historically, sensors, meters, gauges, and instruments have been deployed and used within the vertically integrated context of a single task or system. For example, a zone thermostat triggers the in ow of cold or hot air into specific rooms when the measured air temperature deviates from the target set point by an amount in excess of the guard band; the manifold pressure sensor in a car dictates the engine ignition timing adjustment; the household electric meter is the basis for the monthly utility bill; water temperature, salinity, and turbidity sensors are placed at particular junctures in a river to determine the effects of mixing and runoff; and so on. Examples of specific scenarios are innumerable and incredibly diverse, but they have in common the following: the selection of the measurement device, its placement and role in the encompassing system or process, and the interpretation of the readings it produces are all determined a priori, at design time, and the resulting system is essentially closed—sensor readings are not used outside the system.[1]

This situation has changed dramatically over the past couple of decades owing to the following key factors:

- *Embedded computing.* Until the 1990s, the electronics associated with the analog-to-digital conversion, the rescaling to engineering units, and the associated storage and the data processing dwarfed the size and cost of the transducer used to convert the physical phenomenon to an electrical signal. Consequently, these electronics were shared resources wired to remote sensors. Over the past 20 years, digital electronics have shrunk to a small fraction of their former size and cost, have been integrated directly into the sensor or actuator, and have expanded in function to include quite general processing, storage, and communication capabilities. The

---

[1] In settings in which the transducer is physically and logically distinct from the enclosing system, typified by the Highway Addressable Remote Transducer (HART) for process control and Building Automation and Control Networks (BACnet) for building automation, readings are obtained over a standardized protocol, but their interpretation remains entirely determined by the context, placement, and role of the device in the larger process. The use of the information produced by the physical measurement, and hence its semantics, are contained within the enclosing system.

configurable, self-contained nature of modern instrumentation reduces the costs of deployment and enables broader use.

- *Information-rich operation.* The primary control loop of operational processes (typically represented in manufacturing as plant-sensor-controller-actuator-plant) is usually augmented with substantial secondary instrumentation to permit optimization. For example, in refinery process control, such additional instrumentation streams help to tune controllers to increase yield or reduce harmful by-products. In semiconductor manufacturing, they are employed in conjunction with small-scale process perturbation and large-scale statistical analysis in order to shorten the learning curve and reach a final configuration more quickly. In environmental conditioning for buildings, multiple sensing points are aggregated into zone controllers. Automotive instruments are fused to present real-time mileage information to the driver.
- *Cross-system integration.* Measurements designed for one system are increasingly being exploited to improve the quality or performance of others. For example, light and motion sensors are installed to modulate the amount of artificially supplied lighting in many "green buildings." But those motion detectors are then also available to serve as occupancy indicators in sophisticated heating, ventilation, and air conditioning (HVAC) controls. Rather than simply isolating indoor climate from external factors, modern design practice may seek to exploit passive ventilation, heating, and cooling; to do so requires the instrumentation of building configuration (such as open and closed window and door states) and of external and internal environmental properties (temperature, humidity, wind speed, etc.). All of these sources of information may also be exploited for longitudinal analysis, to drive recommissioning, retrofitting, and refining operations. Interval utility meter readings are used not just for time-of-use pricing but also to guide energy-efficiency measures. Traffic measurements and content-condition instrumentation are applied to optimize logistics operations.

The factors described above have changed the role of instrumentation and measurement from a subsidiary element of the system design process to an integrative, largely independent process of design and provisioning of physical information services. For many sustainability challenges, methodologies are needed that can start with an initial model that is based on modest amounts of data collected during the design process; those methodologies would then include the development of an incremental plan for deploying sensors that progressively improves the model and exploits the improvements to achieve the goals of the system. In many sustainability applications, such as climate modeling and building modeling, the most effective approach may involve combining mechanistic modeling with data-driven modeling. In these applications, mechanistic models can capture (approximately) the main behaviors of the system, which can then be refined by data-driven modeling. Classically, models may be developed from first principles based on the behavior laws of the system of interest, given sufficiently complete knowledge of the design and implementation of the system. Such approaches are reflected not just in the instrumentation plan, but in simulation tools and analysis techniques. However, for most aspects of sustainability, the system may not be rigorously defined or carefully engineered to operate under a narrow set of well-defined behaviors. Examples include watersheds, forests, fisheries, transportation networks, power networks, and cities. New technical opportunities for addressing the challenges presented by such systems as well as opportunities in instrumentation and measurement are emerging, several of which are discussed below.

Coping with Self-Defining Physical Information

Rather than simply drawing its semantics and interpretation from its embedding in a particular system, each physical information service could be used for a variety of purposes outside the context of a particular system and hence should have an unambiguous meaning. The most basic part of this problem is the conversion from readings to physical units and the associated calibration coefficients and correction function.[2] The much more significant part of the problem is capturing the context of the observation that determines its meaning.[3] For example, in a building environment, supply air, return air, chilled water supply, chilled water return, outside air, mixing valve inputs, economizer points, zone set point, guard band, compressor oil, and refrigerated measurement all have physical units of temperature, but these measurements all have completely

---

[2] These aspects have been examined and partially solved over the years with electronic data sheets, such as the IEEE [Institute of Electrical and Electronics Engineers] P1451 family, ISA [Instrumentation Systems and Automation Society] 104 Electronic Device Description Language, or Open Geospatial Consortium Sensor Model Language (SensorML). However, many variations exist within distinct industrial segments and scientific disciplines; the standards tend to be very complex, and adoption is far from universal.

[3] One example of this problem is a stream-water temperature sensor that is normally submerged but under low-water conditions becomes an air-temperature sensor instead. How should this contextual change in semantics be captured? One possibility might be a subsequent data-cleaning step that determines in what "mode" the sensor-context combination was (in this case, perhaps by using a stream-flow sensor or by correlating with a nearby air-temperature sensor). Another example is a soil-moisture sensor whose accuracy can increase with time when more is known about the soil composition—the parameterized equations used by the sensors can be tuned to the soil-type details.

different meanings. The same applies to the collection of measurements across many scientific experiments. Typically, contextual factors are captured on an ad hoc basis in naming conventions for the sense points, the presentation screens for operations consoles, or the labels in data-analysis reports.

The straightforward application of known techniques can be employed to collect the diverse instrumentation sources and deposit readings into a database for a specific setting or experiment. Similarly, electronic records can be made of the contextual information to permit an analysis of the data. The collection, storage, and query-processing infrastructure can be made to scale arbitrarily; processes can be run to validate data integrity and to ensure availability; and visualization tools can be introduced to guide various stakeholders.

To provide these capabilities in general rather than as a result of a design and engineering process for each specific domain or setting, however, requires either significant innovation in the techniques deployed or the development of new techniques. There are, for instance, well-developed techniques for defining the meaning, context, and interpretation of information directly affected by human actions, where these aspects are inherently related to the generation process.[4] To cope with many large-scale sustainability challenges, similar capabilities need to be developed for physical or non-human-generated information.

Closely related to this definitional problem is the family of problems related to registration, lookup, classification, and taxonomy, much as for human-generated information, as one moves from physical documents to interconnected electronic representations. When an application or system is to be constructed on the basis of a certain body of physical information, how is the set of information services discovered? How are they named? If such information is to be stored and retrieved, how should it be classified? If physical information is to be accessed through means outside such classifications, how is it to be searched? Keyword search can potentially apply to the metadata that capture context, type, and role, but what about features of the data stream itself?

Today one addresses these problems by implicitly relying on the enclosing system for which the instrumentation is collected. As physical information is applied more generally, it becomes necessary to represent the model of the enclosing system explicitly if it is to be used to

---

[4]For example, the inventory of products in a retail outlet is quite diverse, but schemas are in place to capture the taxonomy of possible items, locations in the supply chain or in the store, prices, suppliers, and other information. Actions of ordering, shipping, stocking, selling, and so on cause specific changes to be made in the inventory database.

give meaning to the physical instrumentation. However, general model description languages and the like are still in their infancy.

The Design and Capacity Planning of Physical Information Services

Once the physical deployment of the instrumentation capability is decoupled from the design and implementation of the enclosing system, many new research questions arise. Each consumer of physical information may require that information at different timescales and levels of resolution. Furthermore, the necessary level of resolution can change dynamically depending on the purpose of the measurements. In principle, one could measure everything at the finest possible resolution, but this is rarely practical because of limitations in power consumption, local memory, processing capacity, and network bandwidth. What is needed for many sustainability-related challenges is a distributed system by which information needs can be routed to relevant sensors—for the purposes of this discussion, comparatively high bandwidth sensors are meant—and those sensors can then modulate their sampling rates and resolution as necessary.[5]

Recent advances in compressed sensing (to help conserve bandwidth and power) and network coding (to take advantage of network topologies for increasing throughput) add to the complexity of such a distributed system. One can imagine tools that take as input a collection of information consumers, a set of available sensors, and an understood network topology and produce as output a set of sensing and routing procedures that incorporate compressed sensing and network coding. However, this perspective assumes that the locations of the sensors and the network topology are already known. In virtually all practical situations, determining the number, location, and capabilities of individual sensors is an important design step. To support these design decisions, algorithms are needed for sensor placement and sizing. These algorithms require models of the phenomena being measured and of the information needs of each consumer. How will such models be provided and in what representation?

As mentioned above, system architecture has traditionally been organized as a cycle: plant-sensor-controller-actuator-plant. In this model, sensor readings are centralized and aggregated to produce an estimate of the

---

[5]Consider, for example, a thermometer in a freshwater stream. For purposes of hydrological analysis, it might suffice to measure only the daily maximum and minimum temperatures and report them once per week. But suppose that one seeks to detect sudden temperature changes that might indicate the dumping of industrial wastewater. Then the thermometer may need to measure and report temperatures at 15-second intervals.

state of the plant. The controller then determines the appropriate control decisions, which are transmitted to the actuators. However, as the "plant" becomes a large, spatially distributed system (e.g., a city, a power grid, an ecosystem) and the volume of data becomes overwhelming, it is no longer feasible to integrate centrally the state estimation and decision making. A recent International Data Corporation study[6] suggests that there will be more than 35 zettabytes of data stored in 2020. Distributed algorithms are needed that can push as much computation and decision making as possible out to the sensors and actuators so that a smaller amount of data needs to be moved and stored. At the same time, these algorithms will need to avoid losing the advantages of data integration (reduction of uncertainty and improved forecasting and decision making).

Finally, tools are needed to support the planning and design process. These tools need to provide the designer with feedback on such things as the marginal benefit of additional sensing and additional network links, the robustness of the design to future information needs, and so forth. In summary, all aspects of capacity planning present in highly engineered systems, such as data centers and massive Internet services, arise in the context of the physical information service infrastructure.

Software Stacks for Physical Infrastructures

Potential solutions to the problems delineated above suggest that sophisticated model-driven predictive control and integrated cross-system optimization will become commonplace rather than rare. As discussed in Chapter 1, on the electric grid today, the independent service operator attempts to predict future demand and to schedule supply and transmission resources to meet it, with possibly coarse-grained time-of-use rates or, in rare cases, critical peak notification to in uence the demand shape. In the future, environmental control systems for buildings may be able to adapt to the availability of non-dispatchable renewable supplies on the grid, using the thermal storage inherent in a building to "green" the electricity blend and ease the demand profile. Distributed generation and storage may become more common in such a cooperative grid. Various analysis, forecasting, and planning algorithms may operate over the physical information and human-generated information associated with the grid, the building, the retail facility, the manufacturing plant. It becomes important to ask what the execution environment is for such control algorithms and analytical applications. What are the convenient abstractions

---

[6]International Data Corporation, "The 2011 Digital Universe Study: Extracting Value from Chaos" (June 2011), available at http://www.emc.com/collateral/demos/microsites/emc-digital-universe-2011/index.htm.

of physical resources that ease the development of such algorithms, and how is access to shared-resource protected and managed? In effect, what is the operating system of the building, of the grid, of the plant, of the eet, of the watershed? Today these operating systems are rudimentary and consist of ad hoc ensembles of mostly proprietary enterprise resource planning systems, building management systems, databases, communication structures, operations manuals, and manual procedures. An important challenge for computer science research is to develop the systems and design tools that can support effective and exible management of these complex systems.

INFORMATION-INTENSIVE SYSTEMS

Sustainability problems raise many research questions for information-intensive systems because of the nature of the data sources and the sheer amount of data that will be generated.[7] Computer science has applied itself broadly to problems related to discrete forms of human-generated information, including transaction processing, communications, design simulation, scheduling, logistics tracking and optimization, document analysis, financial modeling, and social network structure. Many of these processes result in vast bodies of information, not just from explicit data entry but through implicit information collection as goods and services move through various aspects of the supply chain or as a result of analyses performed on such underlying data. The proliferation of mobile computing devices adds not just new quantities of data, but new kinds of data as well. Some data-intensive processes are extremely high bandwidth event streams, such as clickstreams from millions of web users. In addition, computer science is widely applied to discretized forms of continuous processes, including computational science simulation and modeling, multimedia, and human-computer interfaces. In both regimes, substantial data mining, inference, and machine learning are employed to extract specific insights from a vast body of often low-grade, partially related information.

All of the techniques described above can and must be applied to problems associated with sustainability. Nonetheless, several aspects of sustainability, even in addition to the vast quantities of data that will

---

[7]Given the vast amounts of data expected to be generated in the near future, traditional approaches to managing such amounts of data will not themselves be sustainable. Bandwidth will become a significant barrier, meaning that approaches to computation (such as moving computational resources to the data, or computing on data in real time and then discarding them, or other new techniques), different from simply storing the data and computing on them when necessary, will need to become more widespread.

have to be managed, demand greater innovation, or even wholly new techniques, particularly as ever more unstructured data are generated. To a large extent, these challenges arise from the need to understand and manage complex systems as they exist rather than to engineer systems for a particular behavior.

Big Data

Notions regarding the coming wave of "big data"—the vast amounts of structured and unstructured data created every day, growing larger than traditional tools can cope with—and how science in general must cope with it were articulated in *The Fourth Paradigm: Data-Intensive Scientific Discovery*,[8] which posits an emerging scientific approach, driven by data-intensive problems, as the evolutionary step beyond empiricism, analyses, and simulation. Useful data sets of large size or complicated structure exceed today's capacity to search, validate, analyze, visualize, synthesize, store, and curate the information. The complexity of the transformations that must be applied to render some kinds of observations useful to the scientist or decision maker makes better infrastructure imperative. It is necessary so that one can build on the work of others and so that the population of those with useful insight can expand as data products of successively higher levels of integration and synthesis are constructed.

Unfortunately, there is a growing disparity between available software tools and the ability to leverage those on the scale referred to above. Solutions to many complex systems do not parallelize well, and new tools, algorithms, and likely hybrid systems will be needed. Computer science research is needed to go beyond the embarrassingly parallel problems and to find new approaches, methods, and algorithms for solving these problems. More parallel programming tools, methods, and algorithms are needed to leverage these large-scale systems.[9] Progress in CS is needed in order to move from descriptive views of data (reporting on "What happened, where, how many?") to more predictive views ("What could happen, what will happen next if . . .?"), and finally to more prescriptive

---

[8]Tony Hey, S. Tansley, and K. Tolle (eds.), *The Fourth Paradigm: Data-Intensive Scientific Discovery*, Seattle, Wash.: Microsoft Research (2009).

[9]A 2011 report from the National Research Council elaborates on the challenge of and increasingly urgent need for significant advances in parallel programming methods that are coupled with advances up and down the traditional computing stack—from architecture to applications: National Research Council, *The Future of Computing Performance: Game Over or Next Level?*, Washington, D.C.: The National Academies Press (2011).

approaches ("How can the best outcomes be achieved in the face of variability and uncertainty?").

In order to handle big data, new approaches or improvements will be required in data mining, including clustering, neural networks, anomaly detection, and so on. For example, the smart grid will grow in terms of complexity and uncertainty, especially as renewables are made a more significant element of the energy mix. This increasing complexity will create an increasingly complex system of equations that will need to be solved on a shrinking timescale in order to create secure and dispatchable energy over larger geographies. This challenge implies a need for improvements in computational capabilities to cope with problems ranging from relatively simple $N - 1$ contingency analysis, to $N - x$, to an ability to parallelize the solution to very large systems of sparsely populated matrices and equations that run on high-performance computing systems. In addition, appropriate semantic layers will be needed to bridge the various data sources with a common vocabulary and language, in such a way as to make them more universally applicable.

Heterogeneity of Data

Because sustainability problems involve complex systems, the data relevant to those systems are typically very heterogeneous. Challenges lie not just with huge quantities of data, but stem also from the heterogeneity of their structure and the number of data sets often needed to study a topic. In the management of ecosystems (fisheries, forests, freshwater systems), relevant data sources range from detailed ground-based measurements (catch-and-release surveys, tree core data), to transactional data (fish harvest, timber sales), to hyperspectral and lidar data collected by aircraft and satellites. As discussed above, data for smart buildings include not only energy-consumption and outdoor-weather data, but also data on room occupancy, the state of doors and windows (open or closed), thermostat settings, air ows, HVAC operational parameters, building structure and materials, and so on. Dealing with such diverse forms of information arises, to a lesser degree, in multimodal multiphysics simulations, which typically stitch together multiple major simulation capabilities using specialized adapters and a deep understanding of the algorithms employed in each subsystem. Similar situations arise in data fusion problems and large-scale logistics, such as air-traffic control. But such management of vastly heterogeneous information and processing is typically done on a domain-specific basis. New techniques will be required to do so systematically—say, with transformation and coordination languages to orchestrate the process, or automatic transformation and coordination derived from declarative description of the constituent data and processes.

In some cases, "citizen science" data—such as those provided by bird-watchers (project eBird[10]), gardeners (project BudBurst[11]), and individuals who participate in sport fishing and hunting—may be the only available data about particular regions or events. Heterogeneity extends as well to the provenance, ownership, and control of the data. These data are typically not under the control of a single organization. Access (either one-time-only or ongoing) must be negotiated, and there are important security, privacy, and proprietary data issues. Such authorization and access are generally not transitive, and so new techniques must be developed to manage information ow as information services are composed out of other services.

Coping with the Need for Data Proxies

A persistent challenge in sustainability is the meaningful translation of physical, biological, or social variables into an electric signal. More generally, the data of use with respect to sustainability often do not directly measure the quantities of interest, but instead are surrogates. For example, occupancy in a building may be derived from motion detection, infrared signatures, appliance usage, acoustics, imaging, vibration, disruptions, or other factors, but to varying degrees these may provide only a noisy indication of room occupancy. Ecology focuses on the interactions among organisms (e.g., mating, hatching, predation, infection, defense), but these interactions are virtually never observed directly. An animal (or a disease) may be present in an ecosystem and yet fail to be detected by observations or instruments. Weather radar can detect the movement of birds but not the species or the full direction of motion. Over-the-counter drug sales and web queries can be proxies for the prevalence of u. As another example, measuring snow-water equivalent is normally estimated by sensor pressure at the snow-soil interface. However, snow is not a uid and thus may bridge over a sensor. Other challenges arise from the sensors, which themselves may affect the thing needing to be measured. This may not be chie y an IT problem, but it creates barriers to useful data creation.

Supervised and unsupervised machine learning techniques, as well as those used for gesture recognition, intrusion detection, and preference characterization, may be applied to infer quantities of interest from

---

[10]See http://ebird.org/content/ebird/.
[11]See http://neoninc.org/budburst/.

available surrogates, but the scale and fidelity needed to make substantial headway on sustainability problems require significant transformations of these techniques. Moreover, as these quantities would feed into an extensive process of modeling and automated decision making rather than providing a one-time suggested action, it would be necessary to propagate the uncertainty quantification along with such derived metrics (see below the section "Decision Making Under Uncertainty").

Coping with Biased, Noisy Data

Many data sources are biased (in a statistical sense) and noisy. For example, weather and radar data are collected at special locations (e.g., airports) that were likely chosen to reflect the primary purpose of the data, which may be far from ideal for assessing other topics of interesting, such as climate effects. Carbon-flux towers are usually deployed in at areas where the models that guide instrument interpretation are well understood, but not necessarily representative of important topography. Many citizen scientists make their observations near their place of residence, rather than by following a carefully designed spatiotemporal sampling plan.

Data from sensors can be very noisy. Indeed, the sensor network revolution aims to transition from deploying a small number of expensive, highly accurate sensors to much larger numbers of inexpensive (and less reliable) ones. Moreover, sustainability problems often involve the analysis of longitudinal data (e.g., historical weather records, historical power consumption, historical carbon dioxide concentrations) that have been produced by multiple technical generations of sensors and data-collection protocols. Hence, the data are not of uniform quality. Each new generation of sensor and each change in the protocol may affect the biases and noise properties of the data.

As one significant example, many aspects of climate change center on the increase in global surface temperature. Although there is now broad scientific consensus that human activity is a significant contributor to global climate change, there is much continuing debate about the exact nature of the phenomenon and (more importantly) about the prognosis for the future. Regarding the global surface temperature in particular, Earth's temperature is a very complex phenomenon that varies widely over the surface at any point in time and varies in complex ways over time. There has been assembled over the past hundred years a data set of over 1.6 billion temperature readings at various points over land and seas, using various instruments, by various methodologies, at over 39,000 unique stations.[12] Significant algorithmic work must take place in order to allow the development of a meaningful temperature reconstruction from this complex, noisy, incomplete data set.

Most existing machine learning and data-mining tools make assumptions that are not valid for these kinds of data. They assume that the data are collected at a single temporal and spatial resolution; methods are needed that can work with heterogeneous data. Existing machine learning and predictive data-mining tools typically were designed under the assumption that the data directly measure the desired input-output relationship rather than measuring surrogates. Finally, existing methods assume that the data can be cleaned and rationalized so as to remove noise, impute missing values, and convert multigeneration and multisource data into a common format. However, this process tends to be manual and carried out relative to a particular study of interest by individuals highly trained in the area of study. Even so, it tends to result in a least-common-denominator data set in which numerical, spatial, and temporal resolution are all set to the coarsest level observed in the data. This process may fail to address the different sampling biases and error properties of different data sources, and hence it has the potential to introduce errors into the data.

A fundamental computer science challenge is to automate this cleaning and rationalization process as much as possible and to make it systematic, able to scale to vast streams of continuous data, and able to retain the full information content. A related challenge is to provide tools that support the visual detection of data anomalies and, once an anomaly is found, fast methods for finding and repairing all similar anomalies. An analyst would like to ask, "Are there other data that look like this plot?" and obtain useful answers despite variations in underlying sampling rate, fidelity, and format.

Coping with Multisource Data Streams

Most early machine learning and predictive data-mining tools were designed for the analysis of a single data set under the assumption that the data directly measure the desired input-output relationship; the goal was to learn a mapping from inputs to outputs. Increasingly these techniques

---

[12]The Goddard Institute for Space Studies (GISS) maintains a record of global surface temperatures as well as information on their analysis and on GISS publications, which is available at http://data.giss.nasa.gov/gistemp/. See also National Research Council, *Surface Temperature Reconstructions for the Last 2,000 Years*, Washington, D.C.: The National Academies Press (2006), according to which historical temperature measurements go back to about 1850, and proxy temperature measurements go back millions of years but are most prevalent for the past two millennia.

are applied in real-time settings, such as in massive Internet services both for the adaptive optimization of the complex distributed system providing the service and in adapting the service to optimize the user experience, the profit derived, and so on. One path for CS research in dealing with the reality of noisy, multigeneration, multisource data streams is to develop machine learning and data-mining algorithms that explicitly model the measurement process, including its biases, noise, resolution, and so on, in order to capture the true phenomenon of interest, which is not directly observed. Such methods should treat the data in their original, raw form so that they can capture and take account of different properties of each generation of instrumentation. In some cases, it is possible to focus on empirical model development by applying unsupervised learning methods to extract relationships between inputs and outputs without establishing a specific physical interpretation of the input values. This way of proceeding can sometimes provide insight without the need to isolate the bias, eliminate noise, and calibrate readings.

The field of statistics has studied latent variable models, such as that described in Box 2.1, in which the phenomenon of interest is modeled by one or more latent (hidden) variables (e.g., Z_s). However, existing statistical methods rely on making strong parametric assumptions about the probability distributions governing the latent variables. An important research challenge is to transform these statistical methods into the kinds of exible, non-parametric methods that have been developed in computer science (support vector machines, ensembles of tree models, and so on).[13] Such a transformation should also result in methodologies that are easy to apply by non-statisticians and non-computer-scientists.

The use of these techniques may in some circumstances sharpen the tension between information quality and privacy. If the raw data have been obfuscated to enhance privacy, the latent variable model will seek to undo this and infer the un-obfuscated form. When is it appropriate to do this? Are there effective and broadly applicable ways to preserve privacy and proprietary rights while still applying these methods?

A further challenge raised by these problems is how to validate hidden-variable models. Traditional statistical methods rely on goodness of fit of a highly restricted parametric model; modern machine learning methods rely on having separate holdout data to test the model. Neither of these approaches will work here—at least not in their standard form. One promising direction is to develop simulation methodologies to evaluate the identifiability of the model; another is to develop methods for

---

[13] R.A. Hutchinson, L.P. Liu, and T.G. Dietterich, Incorporating boosted regression trees into ecological latent variable models, *Twenty-Fifth AAAI Conference on Artificial Intelligence*, pp. 1343-1348 (2011).

BOX 2.1
Understanding the Gap Between Observation and Truth

The reality of noisy, multigeneration, multisource data streams requires new machine learning and data-mining algorithms that explicitly model the measurement process, including its biases, noise, and resolution, and so on, in order to capture the true phenomenon of interest, which is not directly observed.[1] Such methods should treat the data in their original, raw form so that they can capture and take account of different properties of each generation of instrumentation.

In general, there is a gap between observation and underlying truth. In a basic sense, this gap exists whenever a transducer is used to measure a phenomenon—in addition to the mapping from input stimulus to output reading, there is a question of how the transducer is coupled to the underlying phenomenon generating the stimulus. How is the accelerometer bonded to the vibrating beam? How does the soil-moisture sensor disturb the hydrological behavior of the surrounding soil? A phenomenological gap may exist regardless of the precision of the sensor.

For example, consider the case of an observer conducting a biodiversity survey. The observer visits a site and fills out a checklist of all of the species observed. The resulting data provide a record of *detections* rather than a record of the true presence or absence of the species. The latter can be captured by an occupancy/detection model,[2] as shown in Figure 2.1.1.

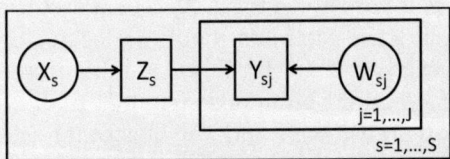

FIGURE 2.1.1 Probabilistic graphical model representation of the occupancy/detection model. Observed variables are shaded; S is the number of sites; J is the number of visits to each site. An observer visits site s at time j and reports $Y_{sj} = 1$ if the species is detected and 0 if not. The hidden variable $Z_s = 1$ if the species is present at site s and 0 otherwise. X_s is a vector of attributes of the site, and the $X \rightarrow Z$ relationship predicts whether, based on these attributes, the species will be present. This is the *occupancy model* that is of primary interest. W_{sj} is a vector of attributes that influence detectability (e.g., the level of expertise of the observer, the density of the foliage, the weather, how much time the observer spent, etc.). The $Z \rightarrow Y \leftarrow W$ relationship represents the *detection model*, which predicts the probability that the species will be detected given that it is actually present (and the probability that it will be falsely reported, given that it is actually absent).

---

[1] S.K. Thompson and G.A.F. Seber, *Adaptive Sampling*, Washington, D.C.: Wiley Interscience (1996) offers researchers in fields ranging from biology to ecology to public health an introduction to adaptive sampling.

[2] D.I. MacKenzie, J.D. Nichols, J.A. Royle, K.H. Pollock, J.E. Hines, and L.L. Bailey, *Occupancy Estimation and Modeling: Inferring Patterns and Dynamics of Species Occurrence*, San Diego, Calif.: Elsevier (2005).

ELEMENTS OF A COMPUTER SCIENCE RESEARCH AGENDA 69

BOX 2.1 continued

In settings in which the observation process can be designed in advance, optimal methods exist for determining the spatial layout of sites and the number of visits to each site.[3] When the data are collected by an uncontrolled observation process (e.g., by citizen scientists) or when the underlying process is poorly understood, more flexible machine learning methods are needed.

---

[3]S.K. Thompson and G.A.F. Seber, *Adaptive Sampling*, Washington, D.C.: Wiley Interscience (1996).

---

visualization that permit the inspection and manipulation of the various components of the model.

Although hidden-variable models can address the challenges of multiple generations of sensors and heterogeneous data sources, they do not solve the problem of biases in sampling. There is a need for new methods that can explicitly capture the differences between the spatiotemporal distribution of the sampling process and the desired spatiotemporal distribution of the model.[14]

An additional challenge is to make all of these methods fast enough for interactive use. Current practice in machine learning modeling harkens back to the days of batch processing. Each iteration of model development and evaluation takes several days, because the data are so voluminous that the management tools, algorithms, and visualization methods require several hours to run. Data volumes will continue to explode as the number of deployed sensors multiplies. Can CS researchers develop integrated data-mining and visualization systems that can support interactive model iteration? Such systems would produce a qualitative change in the sophistication of the models that can be applied and the thoroughness with which they can be evaluated.[15]

---

[14]One promising direction is to build on recent advances in covariate shift methods. See J. Quiñonero-Candela, M. Sugiyama, A. Schwaighofer, and N.D. Lawrence (eds.), *Data Shift in Machine Learning*, Cambridge, Mass.: MIT Press (2009). There is also a growing literature in handling preferential sampling for modeling geostatistical processes: see, for example, P.J. Diggle, R. Menezes, and T. Su, Geostatistical inference under preferential sampling, *Journal of the Royal Statistical Society, Series C*, 59(2) (March 2010).

[15]Such integrated data-mining and interactive model iteration will be critical to transforming the electric grid. M. Ilic, Dynamic monitoring and decision systems for enabling sustainable energy services, *Proceedings of the IEEE*, Vol. 99, pp. 58-79 (2011), offers a possible framework in which distributed models evolve as more information becomes available and

ANALYSIS, MODELING, SIMULATION, AND OPTIMIZATION

One key role of computer science in sustainability is to provide technology for model development. Models permit the extraction of meaningful information from context-dependent, potentially noisy measurements and observations of complex, at best partially engineered, systems in the physical world. Models allow the many interrelated aspects to be decomposed into facets so that progress can be made in a somewhat incremental fashion. Models provide a frame of reference for the many distinct but interrelated streams of information. Computational resources and CS techniques can be brought to bear in several ways: to refine the grid size and time step, to improve the model's representation of processes (making it more complex), and to run the model over multiple scenarios (varying the time period, input values, and so on). This section discusses three interrelated topics: multiscale models, the combining of mechanistic and statistical models, and optimization under uncertainty.

Developing and Using Multiscale Models

Multiscale models, the first of the three interrelated topics, represent processes at multiple temporal and spatial scales. For example, a biological population model might include the agent-level modeling of each organism within a landscape, coupled with the ock-level modeling of group behavior at a regional scale, and population-level modeling at a continental scale. Forest fire models could include models of individual tree growth and burning, coupled with the stand-level and landscape-level modeling of fuel load, coupled with regional models of fire ignition and weather. Global multiscale weather models can operate at low resolution over the entire planet but with higher resolution over regions of interest (e.g., for forecasting hurricanes). In a sense, most sustainability problems arise because behavior at one scale (e.g., energy use in automobiles, land use along migration yways) affects phenomena at very different scales (e.g., global climate change, species extinctions), and those larger-scale phenomena then enter a feedback cycle affecting activities at lesser scales. Multiscale models are important for understanding these sorts of problems.

---

the complexity of the interactions among layers is minimized. The complexity of decision making is internalized to the (groups of) system users instead of burdening the system operator with overwhelming complexity. Iterative learning through binding technical and/or economic information exchange is implemented for learning the interdependencies and aligning them with the objectives of the system as a whole. Significant effort is needed to make such frameworks and system designs more concrete.

One standard approach seeks to capture all phenomena by modeling only at the finest scale in space and time. However, there are both computational and information-theoretic reasons why this approach often fails. Computationally, fine-scale models are extremely expensive to run; multiscale methods allow computational savings while still providing fine-scale predictions. From the perspective of information, large-scale emergent phenomena may not be well modeled by aggregating fine-scale models. Small errors at the fine scale may lead to large errors at larger scales. With limited data available to calibrate fine-scale models (clouds, land surface, human dimensions, etc.), it may not be possible to get a good fit to larger-scale phenomena (global or regional climate change). Hence, it is often preferable, both for computational and representational reasons, to model systems at multiple scales while coupling the models so that they in uence one another.

Homogeneous multiscale models, such as models based on Fourier or wavelet analysis and models that employ adaptive grid sizes (e.g., multiscale meteorology models), are reasonably well understood. However, heterogeneous multiscale models, in which models at different scales employ very different representation and modeling methods, are not as well studied. Methods of upscaling, that is, summarizing fine-scale information at coarser scale (e.g., "parameterizations" in climate models), and methods of downscaling, that is, extending coarse-scale models with fine-scale information, have only recently been developed and still require much additional research. Among the questions to be addressed are these: What is the design space of upscaling and downscaling methods? Are there best practices? Can these be encapsulated in general-purpose modeling packages (e.g., as middleware services)? To the extent that upscaling is primarily performed for computational reasons, can it be automated? How do upscaling and downscaling interact with parallel implementations of the models? In addition, how should mismatches in scale (e.g., mismatches in temporal scale in areas such as coupled ocean-atmospheric models) best be handled?

Combining Statistical and Mechanistic Models

The second of three interrelated modeling challenges is that of combining statistical and mechanistic models. (The former are empirical and the latter are derived from first principles.) In many sustainability settings, some aspects of the problem can be captured by mechanistic models—for example, of physics, chemistry, and so on. However, often it is not tractable to construct models with sufficient fidelity to capture the aspects of the phenomena of interest. For example, consider managing the heating, ventilation, and air conditioning of an office building. Mechanistic models of air ow and heat transport can provide a good first approximation of how a building will behave under different HVAC configurations. However, in actual operation, the outdoor environmental conditions, status of doors and windows (open or closed), position of furnishings, installation of other equipment (e.g., space heaters, digital projectors, vending machines), and number and behavior of occupants can produce very different operational behaviors. Introducing all of these secondary and tertiary factors into a mechanistic model may be very demanding; these factors are typically in ux. Alternatively, as more sensors are incorporated into building environments there may be enough data to fit a statistical model of building behavior—that is, to develop an empirical model. However, such statistical models must in effect rediscover aspects of the physics in the mechanistic models, and this can require an immense amount of data. An attractive alternative is to integrate one or more data-driven modeling components with the mechanistic model. A mechanistic model would be modified by inserting statistically trained components with the goal of these components being to transform the inputs (initial state and forcings), dynamics, and outputs so that the model produces more accurate predictions.

Virtually all existing work on the integrating of statistical models into mechanistic models has taken place outside of computer science. A key challenge is to bring these methods into computer science and generalize and analyze them. Among the research questions are the following: What general-purpose algorithms work well for fitting the statistical components of an integrated model? How can overfitting be detected and prevented (i.e., what are appropriate regularization penalties for such integrated models)? What are good methodologies for evaluating the predictive accuracy of such models? Many algorithms for evaluating mechanistic models employ adaptive meshes; how can statistical methods be integrated with mesh adaptation? The fitting of statistical models typically requires evaluating the mechanistic model hundreds or thousands of times. Running global climate models even once at full resolution can require many days of time on the world's largest supercomputers. Can experimental designs that make efficient use of and minimize expensive model runs be implemented? Can the statistical models be fitted on upscaled versions of the mechanistic models and then downscaled for full-resolution runs? Under what conditions would this work?

Decision Making Under Uncertainty

Actually addressing sustainability problems requires that one move beyond observation and analysis to action. But there is inherent uncertainty every step of the way—in the decision making informed by modeling and simulation, in the measurements, in the modeling and simulation itself, and possibly in the basic characterizations of the factors that comprise the system. In terms of science-based decision making, a central challenge is thus the making of (optimal or at least good) decisions under uncertainty. There are many sources of uncertainty that must be taken into account. First, the scientific understanding of many systems is far from complete, and so many aspects of these systems are unknown. Second, even for those aspects for which good mechanistic models exist, the data needed to specify the boundary conditions with sufficient accuracy are often lacking, especially when human decisions and activity need to be included. Third, the lack of data and the computational cost of running the models often require a coarsening of the scale and the introduction of other approximations. Finally, today's sustainability risks are time-critical, and so just waiting for additional scientific and engineering research in order to address these uncertainties is not an option. Action must be taken even as research continues. Therefore, the choice of actions should also take into account the fact that scientific understanding (and computing power) is expected to improve over time, and future decisions can be made with a better scientific basis.

In effect, uncertainty must be treated as a quantity that persists and is accounted for at every stage. It is a product of measurement, data collection, and modeling, along with the data themselves. It is both an input and an output of the analysis, modeling, and simulation efforts. And finally, decisions must be made based not just on expected outcomes, but also on the uncertainty associated with the various alternatives.[16] There are three areas that pose interesting research challenges for computer science with respect to uncertainty: assessment, representation, and propagation of uncertainty; robust-optimization methods; and models of sequential decision making.

Assessment, Representation, and Propagation of Uncertainty

Many models employed in global climate change, natural resource management, and ecological science are deterministic mechanistic mod-

---

[16]Regarding the challenge of the propagation of uncertainties and, nevertheless, how best to use models to help decision making, a paired set of papers on the climate sensitivity problem appeared in 2007 in *Science*: one was a research article and the other a perspective (G.H. Roe and M.B. Baker, Why is climate sensitivity so unpredictable? *Science* 318(5850):629-632 [2007]; M.R. Allen and D.J. Frame, Call off the quest, *Science* 318(5850):582-583 [2007]). Their main point, in the general context, is that decision making based on models of future scenarios must be adaptive. Another point is that estimating the distribution function of model uncertainty requires knowledge of the distributions of the input data.

els that provide no measures of uncertainty. Recently, there has been substantial interest in assessing the uncertainty in these models. The primary method is to perform Monte Carlo runs in which the model parameters and inputs are varied in order to reflect uncertainty in their values, and the propagation characteristic of the uncertainty is reflected in the degree of variation in the outputs. Although valuable, this does not assess uncertainty due to model-formulation decisions and computational approximations. Research questions include the following: How can one minimize the cost of Monte Carlo uncertainty assessment? For example, can program-analysis methods determine that uncertainty in some sets of parameters interact in well-behaved ways (independently, additively, multiplicatively)? Can convenient and efficient tools be provided for authoring, debugging, and testing alternative modeling choices so that the uncertainty engendered by these choices can be assessed? Can existing tools for automated software diversity—which were developed for software testing and security[17]—be extended to generate model diversity?

Models are often part of a sensor-to-decision-making pipeline in which sensor measurements are cleaned and rationalized, then fed to a set of models that simulate the effects of different policy choices and assess their outcomes. However, it is often the case that uncertainty in one stage (e.g., data cleaning) is not retained and propagated to subsequent stages. Existing scientific workflow tools (e.g., Kepler[18]) do not provide explicit representations of uncertainty or standard ways of propagating uncertainty along such pipelines. Additional work is needed to develop such representations and to provide support for automating the end-to-end assessment of uncertainty. For example, it should be possible to automate end-to-end Monte Carlo uncertainty assessment. One can also imagine extending methods of belief propagation from probabilistic graphical modeling in order to propagate uncertainty along data-analysis pipelines automatically. Coupled with modules for representing policy alternatives and modules for computing objective functions, such workflow tools could provide important support for decision making under uncertainty.

Robust-Optimization Methods

The assessment of explicit uncertainty aims to address the "known unknowns." Classical models of decision making typically involve selecting the actions that maximize the expected utility of the outcomes accord-

---

[17]For a review, see A. Gherbi, R. Charpentier, and M. Couture, Software diversity for future systems security, *CrossTalk: The Journal of Defense Software Engineering* 25(5):10-13 (2011).

[18]For more information on Kepler, see https://kepler-project.org/.

ing to some underlying statistical model. Unfortunately, experience with ecological modeling and environmental policy suggests that there are many "unknown unknowns"—phenomena that are unknown to the modelers and decision makers and therefore not accounted for in the models.[19] One possible safeguard is robust optimization.[20] Rather than treating model parameters as known, this approach assumes instead that the parameters lie within some uncertainty set and optimizes against the worst-case realization within these sets. The size of these uncertainty sets can be varied to measure the loss in the objective function that must be sustained in order to achieve a given degree of robustness. Robust-optimization approaches can greatly improve the ability to sustain significant departures from conditions in the nominal model. Existing robust-optimization methods generally assume that the decision model can be expressed as an optimization problem with a convex structure (e.g., linear or quadratic programs). Robust optimization is sometimes considered overly conservative. Convex constraints over multiple uncertainty sets can be introduced to rule out simultaneous extreme events and reduce the over-conservatism of first-generation robust-optimization methods.[21] An open theoretical question is that of determining the best ways to use data in optimization problems. In some problems in which there are insufficient data, the question becomes one of how to properly incorporate subjective opinion about the data and what the best way is to characterize uncertainty. Another research challenge is to develop robust-optimization methods that are applicable to the kinds of complex nonlinear models that arise in sustainability applications.

Optimal Sequential Decision Making

Most sustainability challenges will not be addressed by a decision made at a single point in time. Instead, decisions must be made iteratively over a long time horizon since a system is not sustainable unless it can be operated indefinitely into the future. For example, in problems involving natural resource management, every year provides a decision-making opportunity. In fisheries, the annual allowable catch for each species must be determined. In forests, the location and method for tree harvesting

---

[19] D.F. Doak et al., Understanding and predicting ecological dynamics: Are major surprises inevitable? *Ecology* 89(4):952-961 (2008).

[20] A. Ben-Tal, L. El Ghaoui, and A. Nemirovski, *Robust Optimization*, Princeton, N.J.: Princeton University Press (2009).

[21] D. Bertsimas and A. Thiele, Robust and data-driven optimization: Modern decision-making under uncertainty, *INFORMS Tutorials in Operations Research: Models, Methods, and Applications for Innovative Decision Making*, pp. 1-39 (2006).

must be specified, as well as other actions such as mechanical thinning to reduce fire risk. In energy generation and distribution, the location of new generation facilities and transmission lines must be chosen. In managing global climate change, the amount of required reduction in greenhouse gas emissions each year must be determined. The state of the art for solving sequential decision problems is to formalize them as Markov decision problems and solve them by means of stochastic dynamic programming. However, an exact solution through these methods is only feasible for processes whose state space is relatively small (tens of thousands of states). Recently, approximate dynamic programming methods have been developed in the fields of machine learning and operations research.[22] These methods typically employ linear function approximation methods to provide a compact representation of the quantities required for stochastic dynamic programming.

An important aspect of sustainability problems is that they often involve optimization over time and space. For example, consider the problem of designing biological reserves to protect threatened and endangered species and ecosystems. Many conflicting factors operate in this problem. Large, contiguous reserves tend to protect many species and preserve biodiversity. However, such reserves are also more vulnerable to spatially autocorrelated threats such as fire, disease, invasive species, and climate change. The optimal design may thus involve a collection of smaller reserves that lie along environmental gradients (elevation, precipitation, etc.). The purchase or preservation of land for reserves costs money, and so a good design should also minimize cost. Another factor is that reserves typically cannot be designed and purchased in a single year. Instead, money becomes available (through government budgets and private donations) and parcels are offered for sale over a period of many years. Finally, the scientific understanding and the effectiveness of previous land purchase decisions can be reassessed each year, and that should be taken into account when making decisions.

The solution of large spatiotemporal sequential decision problems such as those described above is far beyond the state of the art. Striking the right balance between complexity and accuracy, especially in the context of complex networked systems, is critical. New research is needed to develop methods that can capture the spatial structure of the state each year and the spatial transitions (e.g., fire, disease) that occur. There are sustainability problems in which all three of these factors—uncertainty, robustness, and sequential decision making—combine. For example, in

---

[22] W. Powell, *Approximate Dynamic Programming: Solving the Curses of Dimensionality* (2nd Ed.), New York, N.Y.: Wiley (2011).

reserve design, models of suitable habitat for threatened and endangered species are required. These are typically constructed by means of machine learning methods and hence are inherently uncertain. This uncertainty needs to be captured and incorporated into the sequential decision-making process. Finally, existing stochastic dynamic programming methods are designed to maximize expected utility. These methods need to be extended in order to apply robust-optimization methods. A research opportunity is to integrate the training of the machine learning models—which can itself be formulated as a robust-optimization problem—with the robust optimization of the sequential decision problem. This integration would allow the machine learning methods to tailor their predictive accuracy to those regions of time and space that are of greatest importance to the optimization process and could lead to large improvements in the quality of the resulting decisions.

Formulating problems in terms of sequential decision making can sometimes make the problems more tractable. For example, Roe and Baker[23] show that structure inherent in the sensitivity of the climate system makes it extremely difficult to reduce the uncertainties in the estimates of global warming. However, by formulating the problem as a sequential decision-making problem, Allen and Frame[24] show that it is possible to control global warming adaptively without ever precisely determining the level of climate sensitivity.

HUMAN-CENTERED SYSTEMS

It is critical, for real-world applicability, to situate technology innovation and practice within the context-specific needs of the people benefiting from or otherwise affected by that technology. For example, in the context of introducing intelligence into the electric grid in order to increase sustainability, the essential measures and relevant information are very different when considered from the differing perspectives of the utility, supplier, and customer. The utility may be interested in introducing payment schedules that in uence customer behavior in a manner that reduces the need to build plants that run for only a tiny fraction of the time (to serve just the diminishing tail of the demand curve). Avoiding such construction does reduce overall GHG emissions, but the primary goal is to avoid capital investment. Trimming the peak does little to reduce overall energy use, but it reduces the use of the most costly supplies. A consumer-centric perspective is likely to focus on overall energy savings

---

[23]G.H. Roe and M.B. Baker, Why is climate sensitivity so unpredictable? *Science* 318 (5850):629-632 (2007).

[24]M.R. Allen and D.J. Frame, Call off the quest," *Science* 318(5850):582-583 (2007).

and efficiency measures, not just on critical-peak usage. Thus, greater emphasis may be placed on visualizing usage, understanding demand, reducing waste, curbing energy consumption and less important usage, and (if there is dynamic variable pricing) helping to move easily rescheduled uses (e.g., water heating) to off-peak times. A grid-centric perspective, by contrast, may focus on the matching of supply and demand, as well as on the utilization of the key bottlenecks in the transmission and distribution infrastructure. All of these stakeholders need to be considered, and ideally involved, to substantially increase the penetration limit of non-dispatchable renewable supplies, because of the need to match consumption to supply. And, all stakeholders have substantial needs for monitoring usage data, determining causal relationships between activities and usage, and managing those activities to optimize usage. In addition to the needs and values of these direct stakeholders in the technology, the indirect stakeholders should also be considered—that is, those who are affected by the technology but do not use it. In the smart grid example, the set of indirect stakeholders is broad indeed, since everyone is (for example) affected by climate change. The ability to understand such needs and to guide the development of technology on that basis constitutes a natural application of techniques developed in the area of human-computer interaction (HCI).

More generally, a human-centered approach can and should be integrated with each of the topics discussed above. Issues such as human-in-the-loop training of machine learning systems, the interpretability of model results, and the possible use (or abuse) of large volumes of sensed data become particularly salient with a human-centered viewpoint. Indeed, with the vast quantities of data to be generated and used as described earlier, privacy becomes a first-order concern. The role of computer science in sustainability is predicated on the ability to capture and analyze data at a scale without precedent. The understanding and mitigating of privacy implications constitute an area in which fundamental CS research can play a role—in both formalizing the questions in an appropriate way (and indeed this is research well underway) and potentially in providing solutions that can help mitigate the loss of privacy that is, to some extent, inherent in taking full advantage of the power of information-gathering at a global scale with high resolution. It is essential that a human-centered approach be integrated with more traditional security approaches: not only should the techniques for preserving privacy be technically sound, but they should also be accessible, understandable, and convincing to the users of these systems.

Historically, much of the research on sustainability in HCI has focused on individual change. Perhaps one of the best-recognized examples is eco-feedback technology, which leverages persuasive interface techniques[25] and focuses primarily on residential settings. Reduced individual use can socialize people to the issues at hand, and can, at scale, have a direct if limited effect on overall energy use.[26] However, population growth alone may outstrip the gains realized by such approaches. In response, the committee notes the importance of significantly increased attention to social, institutional, governmental, and policy issues in addition to issues of individual change. A challenging public policy question, for example, is how to verify compliance with GHG emissions requirements. Reliably validated carbon reductions, for instance, are important not just to global progress; they would be also invaluable for guiding sustainability efforts at a macro level.[27]

This report emphasizes opportunities for research, in addition to the data and privacy challenges mentioned earlier, on human-centered systems both at the individual level and beyond (at the organizational and societal levels). Examples of such research areas include visualization and user-interaction design for comprehensibility, transparency, legitimation, deliberation, and participation; devices and dashboards for individuals and institutions; expanding the understanding of human behaviors, empowering people to measure, argue for, and change what is happening; and education. Following are brief discussions of each of these.

Supporting Deliberation, Civic Engagement, Education, and Community Action

As noted in Chapter 1, moving toward a more sustainable society will require massive cultural, social, political, and economic changes—and today's technologies are deeply intertwined with many of these changes. Technology can help to support an informed and engaged citizenry. Currently, civic engagement is uneven at best, and thoughtful public deliberation about major issues is often challenging to accomplish. However, the ease of information access, the existence of community-based knowledge

---

[25] For example, see J. Froehlich, L. Findlater, and J. Landay, The design of eco-feedback technology, in *Proceedings of the 28th International Conference on Human Factors in Computing Systems*, New York, N.Y.: Association for Computing Machinery, pp. 1998-2008 (2010).

[26] For a provocative essay on this issue, see P. Dourish, *Print This Paper, Kill a Tree: Environmental Sustainability as a Research Topic for HCI*, LUCI-2009-004, Laboratory for Ubiquitous Computing and Interaction, University of California, Irvine (2009), and a related article: P. Dourish, HCI and environmental sustainability: The politics of design and the design of politics, in *Proceedings of the 2010 ACM Conference on Designing Interactive Systems*, Aarhus, Denmark, pp. 1-10 (2010).

[27] See National Research Council, *Verifying Greenhouse Gas Emissions: Methods to Support International Climate Agreements*, Washington, D.C.: The National Academies Press (2010).

repositories, and the search and social networking capabilities online are transforming the manner in which humans learn, make decisions, and interact. These techniques can be adapted for a research program on designing, deploying, and testing innovative ways for citizens to deliberate and to engage with government and one another, particularly with those who may hold very different views in the context of sustainability. These deliberations should be closely coupled to data (gathered both by professional scientists and citizen-scientists) and simulation results—affordances should be provided to help ground the discussion in the scientific evidence. Similarly, online curricula for students in kindergarten through grade 12 and for adults can explore, for instance, ongoing scientific and policy discussions related to sustainability; and educational initiatives can contribute to societal changes needed to meet sustainability goals.

In addition to opportunities with respect to tools for engaged citizens generally, there are also promising areas of research in helping scientists provide more effective input into these broader discussions and debates on sustainability and potential initiatives. The intellectual merit of this research would center on the issues of how to facilitate large-scale online deliberation about contentious issues; the broader impacts would be in making the results of scientific inquiry more widely seen and discussed. As an example, suppose that there was a network supporting online deliberation among scientists concerned with sustainability for developing key points, areas of strong consensus, areas of disagreement, and supporting evidence. Those deliberations would produce a sustainability action agenda that could be introduced to the public by means of interesting interactive environments designed to appeal to those of all ages. These sites could feed information by means of different media outlets (both traditional and emerging) as well as providing interactive scenarios that people could use to answer questions and debate solutions. One highlight of this system would be a series of consensus news stories, perhaps on a weekly basis. These stories could be based on agenda items created by scientists and rated by public interest.

A core component of such a public education system could be a forum for discussing scientific data, for voicing views on which stories to present and when, and for suggesting how to frame them (deliberative forums for the science community for building consensus positions). A key research issue here is the development of technologies that help organize the discussion, both for long-standing participants and for people who are interested in entering into a long-running discussion but could use help in understanding it and in making useful contributions. The forum should include affordances that make it easy and natural to classify suggestions, pro and con arguments, and so on, to keep this type of exchange from degenerating into just a free-form discussion board. Another important kind of affordance would be hooks for giving sources for assertions (tools to encourage grounding arguments in the scientific data).

Another project might be a highly visible forum for exchanges between groups with quite divergent views in a deliberative setting. Again, the system should include affordances that make it easy and natural to classify arguments and perspectives and tools that encourage the grounding of arguments in the scientific data.

Basic research in educational technology is also crucial to increasing the relevance and effectiveness of tools for a culturally and economically diverse population. Arguably, better support for deliberation and engagement will not be enough. Supporting community action is also essential. In recent years technology has become more and more salient as an enabler of successful social change.[28] In another example, on a local scale, citizen sensing of environmental indicators (e.g., pollutants) has influenced the ability of individuals to advocate for change. As the cost of sophisticated sensors comes down, one can expect to see more and more of them employed by end users. A citizenry that engages with and helps to track this information is important to progress on the issues at stake, and this engagement leads to increased education and engagement in addition to increasing the amount of information available in crucial areas. However, this raises fundamental research problems ranging from the creation of these sensors to our ability to use the data effectively despite the inherent uncertainties that arise from its production.

Design for Sustainability

Techniques developed to design for manufacturing, design for mass customization, and user-centric design can expand on the understanding of what it means to design for sustainability. Techniques such as ENERGY STAR ratings for appliances and Leadership in Energy and Environmental Design (LEED) ratings for buildings have had some success in reorienting industry providers and consumers alike toward more sustainable practices. These efforts can be substantially informed by the measurement, information-collection, and model-development techniques described earlier, but can also use HCI techniques for appropriation, reuse, and end-to-end design for technology products. This research can be expanded to shed light on process, distribution, middleware, and other aspects of the production and distribution of products. Technological advances can

---

[28]T. Hirsch and J. Henry, TXTmob: Text messaging for protest swarms, in *Extended Abstracts on Human Factors in Computing Systems,* New York, N.Y.: Association for Computing Machinery, pp. 1455-1458 (2005). DOI: http://doi.acm.org/10.1145/1056808.1056940.

also contribute to the tracking, monitoring, and analysis of the source materials, production processes, distribution, and eventual disposal of products. This information in turn can help to inform purchase decisions, provide better accounting, and otherwise improve the sustainability of the consumer economy.

Human Understanding of Sensing, Modeling, and Simulation

As the availability of sophisticated sensors, information collection, modeling, and dissemination increase, techniques need to be developed to provide in meaningful forms rich, highly disaggregate information to households, small groups, and organizations regarding resource usage (e.g., for electricity or water consumption). In addition to supporting improved decision making about energy use at the organizational and individual levels, this information could provide civil and environmental engineers with a picture at a new level of detail about how and why these resources are being consumed, allowing their science and practice to advance. At the same time, this possibility raises challenging research questions regarding appropriate amounts of information, how to deal with the inherent uncertainties in the data, techniques for evaluating such systems, coupling with other systems on the supply side (e.g., the smart grid), and important value questions regarding fairness, representativeness, security, and privacy. Better data can also drive modeling and simulation, which can help with such activities as predicting important trends, assessing how well proposed policies would meet objectives, and optimizing resource use. Modeling climate change is an obvious example, but there are many others, including a simulation of the evolution of urban areas, freight transport, and natural environments such as forests or rivers. However, to be effective and relevant to policy making and decision making, such modeling work must include careful consideration of how it integrates with deliberation and the political process. This raises issues of design for transparency, legitimation, appropriability, and participation.

Tools to Help Organizations and Individuals Engage in More Sustainable Behaviors

Another area for research concerns tools that make it easier and perhaps even enjoyable for people to engage in more sustainable behaviors. Some of the many examples in this area are the providing of real-time public transit arrival and route information (particularly on mobile devices), online ride-share matching, geowikis for bicycling, Zipcar, Freecycle, and the like. Another class of tools provides eco-feedback: targeted information about resource consumption (perhaps in real time), integrated with suitable visualization techniques and appropriate persuasive technology, for example to show progress toward personal or group goals. This area is also related to the previous opportunity regarding the use of information from resource-usage sensing. It is important to recognize the limits of these technologies: better transit information is great if a good underlying transit system exists, but it is not so useful without that. Similarly, eco-feedback regarding energy use can be helpful, but it does not address the more fundamental, underlying energy challenges in some situations—such as low-income households in which comparatively expensive upgrades would be a financial hardship, or homes that contain inefficient appliances or poor insulation. For such challenges, alternative solutions would be needed.

Many of the techniques described here are relevant to organizations as well. For example, a large organization might similarly provide targeted information about resource consumption, in real time, to show progress toward goals for different branches of the organization.

Mitigation, Adaptation, and Disaster Response

Even under optimistic climate change scenarios, weather disasters are likely to increase in number and severity, resulting in both the need for immediate disaster relief and likely the need to assist large numbers of refugees (e.g., from low-lying regions).[29] Also, unfortunately, human actions are likely to continue to contribute directly to environmental disasters such as oil spills. There are research challenges with respect to developing plans that can be revised rapidly under conditions of great uncertainty, making use of vast numbers of citizen observations (such as micro-content posted from disaster areas by individuals), coordinating supply efforts, and others. One challenge for this line of work is recognizing that there are huge uncertainties about the future and thus also in developing tools and infrastructure that are exible, adaptable, and appropriate.

Using Information from Resource-Usage Sensing

Recent work has opened the possibility of providing rich, highly disaggregate information to households, small groups, and organizations regarding resource usage. For example, immediate feedback can now be

---

[29]National Research Council, *Adapting to the Impacts of Climate Change*, Washington, D.C.: The National Academies Press (2010).

provided on electrical energy use at the appliance or individual lighting circuit level. A number of possibilities arise as a result, including detailed eco-feedback about usage and tighter coupling with smart grid technology on the supply side. Similar feedback is possible for other resources such as water and natural gas.

This possibility does, however, raise a number of challenging research questions. For example, what is the appropriate amount of information to provide to households? Clearly there is the possibility of overwhelming them with information. How are the inherent uncertainties in the data to be dealt with? How are such systems to be evaluated? The traditional HCI evaluation techniques of laboratory studies and small-scale deployments are inadequate, but massive deployments over long periods are slow and expensive, implying that one can only try a small number of alternatives (in tension with the need for rapid prototyping and iteration). How can these systems be coupled with smart grid technology on the supply side? For example, the grid could signal to the household that the system was close to capacity and that lowering energy use for the next hour would be very helpful (or perhaps would result in a lower bill); or, conversely, the household could be signaled that this would be an opportunity for some non-time-critical activity. This arrangement would be a combination of automated actions, with the scripts under the household's control, and explicit actions.

Another set of issues concerns fairness and representativeness. For example, the majority of households in the United States are low-income and many households rent, although most work in this area focuses on relatively af uent homeowners. Can systems and policies be designed that do not unfairly disadvantage some households, particularly ones that can least afford additional charges? Another set of challenges concerns security and privacy. Such systems offer the potential for reducing resource consumption and making better use of resources, but there are clear security and privacy risks if the system is compromised. Related to that issue are questions of responsibility and power around available infrastructure that must be addressed. Not everyone owns a home or pays for energy use, and the relationships between landlords, residents, laws (incentives, disincentives, and so on), available services (green contractors), and other factors in uence energy use outcomes and may bear on the design of technology (for example, in terms of authenticating who has access to what data).

It is difficult to get good information about the fine-grained use of energy right now. Buildings are not generally instrumented to produce these data, yet a true understanding of the forces driving energy use is impossible without better data. Better information about which appliances are in use and when they are in use can help in developing a more complete understanding of human behavior, and perhaps in identifing interventions that can have an impact on energy use. Even a modest advance such as analysis based on the segmentation of a building's energy use among HVAC, lighting, and plug load could yield useful results. Although this may seem like a pure sensing problem, the process of deploying sensors, labeling data, and interpreting the results involves people, and computer science researchers are at the forefront of some of the innovations in this area.[30] Despite these advances, the problem of labeling data and interpreting the results is one that requires more attention.

CONCLUSION

This chapter provides examples of important technical research areas and outlines a broad research agenda for computer science and sustainability. Although there are numerous opportunities to apply well-understood technologies and techniques to sustainability, there are also hard problems—such as mitigating climate change—for which current methods offer at-best partial solutions, and rapid innovation is essential in light of the pressing nature of the challenges. The areas highlighted in this chapter—measurement and instrumentation; information-intensive systems; analysis, modeling, and simulation; optimization; and human-centered systems—are counterparts to well-established research areas in computer science. This overlap has clear positive implications. However, finding a way to have a significant impact may require new approaches to these problems and almost certainly new ways of conducting and managing research. Chapter 3 explores ways of conducting and managing research so that computer science research can have an even greater impact on sustainability challenges.

---

[30]For example, Patel and others have developed comparatively lightweight methods to acquire reasonably fine-grained data in homes; see J. Froehlich, E. Larson, S. Gupta, G. Cohn, M. Reynolds, and S.N. Patel, Disaggregated end-use energy sensing for the smart grid, *IEEE Pervasive Computing, Special Issue on Smart Energy Systems*, January-March (2011).

3

Programmatic and Institutional Opportunities to Enhance Computer Science Research for Sustainability

The challenges of achieving a sustainable society are truly global, with complex interdependencies that affect risk assessments, technical and social opportunities for solutions, and economic and political feasibility. No one field or discipline on its own could possibly be expected to "solve" even one aspect of this problem. However, information is central to making progress on many fronts. Thus computer science (CS)—which couples information and innovation—is vital to sustainability. For computer science to play its part in meeting global sustainability challenges, priority should be given to research that addresses one or more important sustainability challenges (examples were described in Chapter 1) and that offers significant impact. This impact may be direct, or it may be through game-changing contributions that offer significant leveraging opportunities for other domains. In either case, priority should be placed on opportunities to address the sustainability challenge to a tangible degree.

This chapter explores some of the potential impediments within the field of computer science to making significant progress on issues pertinent to sustainability. It considers how to bridge the gap between the traditional research quest for universality[1] and the imperative to have a specific impact on sustainability challenges. The chapter is aimed pri-

---

[1]The committee uses the term "universality" to encompass the related notions of generalizability (solutions that are amenable to relatively straightforward abstractions in order to address more general versions of a given problem) and breadth (solutions that can be revised to be applicable to broad problem domains and spaces). The quest for universality captures the traditional CS research goals of abstractability and broad applicability.

marily at the CS research community—including both researchers and funders. First, it discusses some of the fundamental aspects of CS research and computational thinking and how these aspects are also critical to the sustainability problem space. Then it explores the challenge of universality and emphasizes that a bottom-up approach is not only necessary in the sustainability space but also has precedent in many other areas of deep computer science. It describes the connection between universality, bottom-up approaches, and sustainability. It then offers suggestions on how to structure research to promote meaningful impact on sustainability. Finally, the chapter identifies methodological opportunities for optimizing research outcomes and impacts.

COMPUTER SCIENCE APPROACHES FOR ADDRESSING SUSTAINABILITY

Chapter 2 highlighted the centrality of data and information to sustainability. Given this centrality, computer science and information technology (IT) are essential to meeting sustainability challenges. The challenge for IT experts and CS researchers is in ensuring that technologies and approaches represent usable, appropriate solutions; that they are highly effective; and that they take advantage of the deepest and most powerful insights that can be brought to bear. IT has been and continues to be a critical enabler of progress in vast arenas of society. Sustainability is no exception: IT offers a powerful tool to assist in addressing sustainability challenges.

Moreover, fundamentals of the computer science field itself offer unique and important contributions to sustainability. To name just a few such fundamentals, consider abstraction design, algorithms, operating systems and layering, real-time systems, machine learning, human computer interaction (HCI), and databases. For instance, the very notion of queryable structured data is at the heart of much of computer science; at the same time strides are being made to cope with the vast amounts of unstructured data now available. Given the scope and scale of sustainability challenges along with the vast amounts of relevant data, the structuring and understanding of these data present many challenges. The lens of computational thinking is essential to solving many complex problems,[2] and there are key opportunities within computer science that are clearly

---

[2] See National Research Council, *Report of a Workshop on the Scope and Nature of Computational Thinking*, Washington, D.C.: The National Academies Press (2010); and National Research Council, *Report of a Workshop on the Pedagogical Aspects of Computational Thinking*, Washington, D.C.: The National Academies Press (2011).

> **BOX 3.1**
> **Additional Areas of Promising Computer Science and Related Research for Sustainability**
>
> In addition to the research areas discussed in Chapter 2 of this report, following is a list presenting a sampling of topics and areas that arise in computer science and information technology more generally that are likely opportunities for making progress in sustainability.
>
> - *Science of resilience and adaptive systems.* This is a likely area of opportunity especially as it applies to self-regulating processes, biodiversity, and metrics of adaptability.
> - *Design for robustness, resilience, graceful degradation, and the decoupling of abstractions from implementation* (for instance, designing for average-use cases and building in techniques for degradation, as opposed to designing for peaks with safety margin).
> - *Mass customization, especially in the role of programming languages.* This is a likely area of opportunity for many levels of programming, at many stages of the life cycle.
> - *Understanding technology in context.* One cannot understand how technology will affect sustainability without understanding what people will do with it. The emphasis in computer science on extensibility in system design takes into account the fact that technology as used matters, not just as designed.
> - *Design thinking.* This area is involved with the invention of things that people will use and engage with, which is crosscutting for multiple domains.
> - *Search—a profound advance resulting from decades of research and innovation in multiple areas.* What is the equivalent of search in the physical world? How do we deal with unstructured search, taxonomy, structured query processing—search for data relevant to scientific discovery?
> - *Computer vision.* This field offers likely opportunity as a modality for searching and understanding the physical world.
> - *Representation for purposes of discovery.* This is an area of opportunity in terms of representation of the physical world and of sustainability problems.
> - *Social media.* This area of opportunity relates to information support and sharing, building community, structured argumentation, sensing, modeling, and observation.
> - *Tools for the automated design of very large scale systems.* This is an area of opportunity that includes the development of capabilities to cope with challenges where there is functional decomposition

applicable, even beyond those highlighted in Chapter 2. A sampling of these areas is outlined in Box 3.1.

As one example, many sustainability challenges, particularly those related to infrastructure, make salient the importance of architecture. Architecture encompasses not just structural connections among subsystems, but expectations regarding what a system will do, how its performance will scale, what behaviors are within bounds, and how subsystems (or external actors) should interact with the system as a whole. This type of challenge can be seen perhaps most clearly in the smart grid example discussed in Chapter 1. However, the other illustrative examples in Chapter 1—food systems and the development of sustainable and resilient infrastructure—also make clear that early architectural choices (such as the expected role and behavior of individual farms in the global food system, or the anticipated communications capabilities of first responders in a crisis) can have long-lasting repercussions.

A system's architecture instantiates early design decisions and has a significant effect on the uses, behaviors, and effects of the system long past the time when those decisions were made. Moreover, requirements inevitably change over time, necessitating exible or evolvable designs. Because of this large effect of a system's architecture on almost all aspects of the system over its life cycle, the architecture of larger-scale systems of necessity merits significant attention and resources. As systems have become global in scale, the disciplines of computer science and software engineering have grappled with the challenges of architecture as they pertain to large-scale systems working over large geographic areas, with countless inputs and millions of users. Lessons from architecting hardware, software, network, and information systems thus have broader applicability to the processes of structuring, designing, maintaining, updating, and evolving of infrastructure in pursuit of sustainability.[3]

One question not yet addressed in this report is how well the CS research community is poised to play its part in meeting global sustain-

---

[3] For an in-depth examination of the importance of architecture in software-intensive systems, see Chapter 3 of the following report: National Research Council, *Critical Code: Software Producibility for Defense*, Washington, D.C.: The National Academies Press (2010). It describes the importance of architecture as follows (pp. 68-69): "Architecture represents the earliest and often most important design decisions—those that are the hardest to change and the most critical to get right. Architecture makes it possible to structure requirements based on an understanding of what is actually possible from an engineering standpoint—and what is infeasible in the present state of technology. It provides a mechanism for communications among the stakeholders, including the infrastructure providers, and managers of other systems with requirements for interoperation. It is also the first design artifact that addresses the so-called non-functional attributes, such as performance, modifiability, reliability, and security that in turn drive the ultimate quality and capability of the system. Architecture is an important enabler of reuse and the key to system evolution, enabling management of future uncertainty. In this regard, architecture is the primary determiner of modularity and thus the nature and degree to which multiple design decisions can be decoupled from each other. Thus, when there are areas of likely or potential change, whether it be in system functionality, performance, infrastructure, or other areas, architecture decisions can be made to encapsulate them and so increase the extent to which the overall engineering activity is insulated from the uncertainties associated with these localized changes."

ability challenges. In the committee's view, one perceived barrier to its doing so is the sense that aiming research toward sustainability challenges may conflict with ultimate scientific aims of universality.[4] This question is explored below in more detail.

The most powerful and important computer science innovations to date share the characteristic of universality. Indeed, it is their utility across a wide range of domains that makes their aggregate impact so great. See Box 3.2 for a short list of just some of computer science's most significant achievements. Universities, research laboratories, departments, and funding agencies increasingly recognize the value of multidisciplinary research. However, as the field has matured, there has been comparatively less emphasis on domain-driven approaches to innovation in favor of research that attempts to go directly to universality—that is, abstractability and breadth. Nevertheless, it is a strength of computer science that the field can, and does, ground its advances in real-world problems. As described in Chapters 1 and 2, CS can contribute significantly and critically to sustainability. In this section it is argued that to have the biggest impact on the pressing challenges facing the world today, CS research must be informed with deep knowledge, input, and context from domain experts.

In some areas of computer science, universality is built into the problem definition. Much of theoretical computer science, of course, begins by representing the target problem in abstract, symbolic language. Other examples of research with universality as the focus from the start include the von Neumann computer model itself, programming languages such as Fortran and ALGOL, and early human-factors research (such as that of Doug Englebart and Alan Kay) that created new modes of human-computer interaction. In other equally consequential areas, however, broad applicability has only emerged years or even decades later, as researchers began with domain-specific problems and developed solutions and then later were able to generalize and understand deeper truths from this panoply of specific contributions. Examples of important contributions that began as highly specific projects include the World Wide Web (originally conceived as a means to share research papers and scientific information) and object-oriented programming (early object-oriented languages were developed to address specific problems such as discrete event simulation or graphical interaction).

In this chapter it is argued that CS research on sustainability is best approached from the bottom up: that is, by developing well-structured

---

[4]The integration of computer science with domain sciences was a central tenet of a 1992 National Research Council report: National Research Council, *Computing the Future: A Broader Agenda for Computer Science and Engineering,* Juris Hartmanis and Herbert Lin (eds.), Washington, D.C.: National Academy Press (1992).

BOX 3.2
Universality and Computer Science's Greatest Achievements

The list below offers one view of some of computer science's most significant achievements over the years. "Computer science" is construed broadly here, to encompass information and communications technologies.[1] Most of the achievements listed below were accomplished by focusing on developing domain-specific solutions with an eye toward eventual abstraction and universality.

- Universality—the Turing machine
- Computability and hardness—P, NP, and complexity classes
- Stored program computer—execute anything
- Operating system (versus operator)
- The Internet and Internet Protocol—move anything anywhere
- File storage—move anything forward in time
- Coding, decoding, and cryptography
- Programming language and its translation
- Layered design (versus vertical integration)—abstraction as a foundation
- Microprocessor, the personal computer
- Audio/video representation
- Distributed systems
- Self-describing schemas, documents
- Computer-aided design, computer-aided manufacturing, computer-aided optimization, computer-aided engineering (design, simulate, build, test, measure, use)
- Database query languages and management systems
- Continuous user input (pointers, clickers)
- Graphical user interfaces in various forms
- Parallel execution
- Numerical simulation of physical phenomena
- The web, hypertext, and distributed markup in a global namespace
- Massive keyword search
- Spreadsheets
- Scheduling, planning, optimization
- Very-large-scale integration, design rules, synthesis, verification, massive systems production
- Search techniques, heuristics
- Machine learning
- Structure of graphs

---

[1]There are, of course, other takes on this question. The Computer Science and Telecommunications Board's (CSTB's) well-known "tire tracks" diagram (as seen in National Research Council, *Innovation in Information Technology*, Washington, D.C.: The National Academies Press [2003]) explores innovations in computer science and information technology from the perspective of spawning billion-dollar industries. That list includes time-sharing, client/server computing, entertainment, Internet, local area networks, workstations, graphical user interfaces, very-large-scale integration design, and reduced instruction set computing processors. A CSTB report published in 2004 articulated the essential character of computer science, and focused on seven key themes: computer science (1) involves symbols and their manipulation, (2) involves the creation and manipulation of abstractions, (3) creates and studies algorithms, (4) creates artificial constructs, (5) exploits and addresses exponential growth, (6) seeks the fundamental limits on what can be computed, and (7) focuses on the complex, analytic, rational action that is associated with human intelligence. (National Research Council, *Computer Science: Reflections on the Field, Reflections from the Field*, Washington, D.C.: The National Academies Press [2004].) The differences in these lists are less important than noting the power of focused problem solving in a field whose core strengths include abstraction and adaptability.

solutions to particular, critical problems in sustainability and later seeking to generalize these solutions, as opposed to striving for universality from the start. Many advances will require CS research for progress, as described earlier, but those advances may not be immediately evident as universal approaches. The committee believes that demanding evidence of clear universality from the start is likely to inhibit close interdisciplinary collaboration and, ultimately, major advances. Moreover, many sustainability challenges need to be addressed sooner rather than later, even if imperfectly. At the same time, the fact that the problems associated with sustainability are complex, multifaceted, and in some cases poorly defined means that close attention to experimental robustness and underlying mathematical rigor will be essential.

Previous chapters illustrate this overall approach. For example, the hypothesized research on the smart grid takes an approach to the problem that is fundamentally a computer networks perspective (inspired by the success of the Internet), but it is not initially intended to make universal contributions to the theory of networks. Rather, by focusing on the problem at hand—that of integrated control of power generation, distribution, and use—there is the potential for breakthrough advances on this critical issue at the same time that new computational techniques are being developed. There are clearly general lessons to be learned about these issues. The committee suggests that the best way to learn them is to start with the particular sustainability challenge at hand, make progress on that, and later seek to generalize.

This approach does not mean, however, that any application of computation or IT to problems in sustainability should automatically be seen as computer science *research* for sustainability. Rather, to be judged as a significant contribution within the intersection of CS research and sustainability, the contribution first must have the potential to make a real difference in moving toward a more sustainable future. Second, the contribution must have the potential, if it is successful, to add to generalizable knowledge about sustainability, and the contribution or proposed solution should, at the same time, require new computational techniques or thinking beyond the current state of the art in computing.

The specific criteria for judging research success should of course evolve over time, with members of the community themselves proposing and debating what constitutes the most worthy research. The committee emphasizes, however, the criterion of having the potential to make a real difference. An open research question in its own right is how best to assess and evaluate impacts and how to isolate the effects of any particular sets of interventions, given the scales and time frames of many sustainability challenges.

PRINCIPLE: Encourage research at and across disciplinary boundaries, well informed by specifics and well structured to handle scale, data, integration, architecture, simulation, optimization, iteration, and human and systems aspects. CS research in sustainability should be an interdisciplinary effort, with experts in the various fields of sustainability being equal partners in the research.

TOWARD UNIVERSALITY

Although the committee emphasizes that a premature focus on universality would be detrimental to the kind of high-impact sustainability solutions so desperately needed, universality should not be ignored. Indeed, domain-specific research can lead toward universality. A challenge, though, is how to pursue the universality that contributes so much to the power of computer science. In this section, it is argued that the purposes of domain specificity and contextualization are not at odds with ultimately producing universality in results, and that universality is not achieved directly in most cases in any event. Consider the development of important advances in CS and IT. Achieving universality typically involves developing well-structured innovative solutions, applying them to the problem at hand, evaluating their efficacy, and using this evaluation to guide further improvement, enhancement, and new directions. Successful approaches are then refined and applied in other areas, perhaps similar to the original problem domain, perhaps more remote. As the iterations of application proceed, the universality of the approach is discovered and refined.

Why has this approach worked so well in computer science? Despite the fact that computer science has information at its heart, tools and methods are ultimately instantiated in software. Software is malleable and well disposed to iteration. Software technology is developed, deployed, used, and modified in continuous iterative cycles. Developing modern software is not done through implementing a perfect software system once, at the start. Instead, the state of the art in software engineering urges iteration and architectural flexibility. Software is designed to be updated on a frequent basis over its entire life cycle.

This approach to the creation of software systems has developed for many reasons. For one reason, the work required to discover all or even most of the bugs before release in non-critical systems far exceeds the value of that approach. Similarly, feature sets are expanded through use. The range and number of possible features of any particular target system are larger than what is implemented—if that were not the case, the systems would be even more complex and difficult to use and would take even longer to roll out. Thus systems are rolled out with a modest feature set, and new features are then added over time. This approach allows products to get to market sooner and helps to avoid or reduce the development of features that are not really needed, as well as revealing what actually works in practice. In short, after initial deployment, reality (instead of anticipated or modeled reality) guides the evolution of the feature set. The understanding of the human systems into which all computing systems are deployed is highly limited (not least because the understanding and modeling of humans and organizations are highly incomplete and awed at best). Thus one cannot generally anticipate and simulate all uses. The world of computer systems has grown to evolve features that are adapted to what the users of those systems demand. A virtuous cycle has resulted—users have come to expect exibility and malleability, thus ensuring that feedback loops occur. Systems have versions that are rolled out as available and as features are demanded. Change management is a basic fabric of these systems, which are designed for ongoing change. These systems have innovation and expectations of innovation literally encoded within them.

Another way to think about the inherent adaptability of computer science is to consider an "end to end" argument for the inclusion of authentic applications in systems research. The original end-to-end argument put forward by Saltzer, Reed, and Clark was as follows:

> [F]unctions placed at low levels of a system may be redundant or of little value when compared with the cost of providing them at that low level. . . . The argument appeals to application requirements, and provides a rationale for moving function upward in a layered system, closer to the application that uses the function.[5]

This same logic has implications for systems research and innovation. Authentic applications should be included as part of systems research exploration at as high a level as possible in order to keep functional and performance requirements on a purposeful track. For the purposes of this report, notions of authenticity must encompass high-impact applicability to sustainability challenges.

These fundamentals in computer science are relevant to the way that CS research is done and the way in which research and development investment is approached. This "built for change" characteristic also facilitates the transition from one application to another. Algorithms and their instantiations can be adaptive, iteratively modified to fit a new

---

[5]J.H. Saltzer, D.P. Reed, and D.D. Clark, "End-to-End Arguments in System Design," *Proceedings of the Second International Conference on Distributed Computing Systems*, Paris, France, April 8-10, 1981.

context. More importantly, such iterations are not a significant departure from what occurred in their initial creation; the expectation of iteration is part of the core of the technology. Building for change is done through modularity, through system designs resulting from hard thinking about where to place functionality, through the isolation of errors and details. Change to software happens as the software is developed, and as it is deployed, debugged, and iteratively improved; and it happens as it is applied to a new problem. As a given technique is applied to a new problem, and yet another new problem, and so on, the universality of the technique emerges. For each new application, the characteristics of generality are exposed, and the possibility for further abstraction and broad applicability grows. In the best case, an "exponential" process emerges in which techniques that are broadly applicable are exposed as each successful reapplication enables multiple new adaptations to come to light. Not all potential new applications are developed, of course, but those that are find their ultimate universality through bottom-up cycles of change and through the iterative process of design which promotes that process. Past successful examples of this approach include language translation, Internet protocols, machine learning, object-oriented languages, and databases.

The approaches discussed in Chapter 2 were not described in their most general terms. The committee does not suggest that they be pursued generally. But universality is often seen as the ultimate win of computer science techniques. Although universality is important and must be the goal because some big wins are needed in order to attack the unprecedented challenges of sustainability, the challenges should be approached through the concrete. There are opportunities for CS research to take on the key challenges in sustainability, learn about them, and design focused solutions that work. The design of those solutions should embed the best of CS design and systems learnings—modularity, isolation, simplicity, and so on. Then CS researchers and practitioners should experiment with, apply, and pilot solutions to specific problems; look for the successes and reapply and adapt them to other applications; and develop universality while seeking to increase applicability and impact. If the concrete is embraced across the range of infrastructure, ecosystems, and human systems, reality will help hone and filter possible approaches, and multiple and adapted applications will emerge.

> **FINDING: Fast-moving iterative, incrementally evolving approaches to problem solving in computer science, which were critical to building the Internet and web search engines, will be useful in solving sustainability challenges.**

EDUCATION AND PROGRAMMATICS

The vision described above implies a broadening of what it means to be a computer scientist. A significant opportunity for change is in the area of education. This change should include educating computer science students to achieve impact with computing, computational methods, and systems approaches in important domain-specific areas. Such a shift in culture would encourage these students to develop domain expertise and to collaborate directly with domain experts while in graduate school or in preparing for graduate work[6] and to address such topics as modeling and predicting energy use and designing for reuse.

Making such a shift successfully will also require a culture of experimentation and innovation in the application of computer science. Further, it will require a research infrastructure in order to make progress. That infrastructure should include the following: (1) available standard data sets, models, and challenge problems to the community in order to assist in developing a common discourse and target for innovation, analogous to Grand Challenges in robotics, speech, vision, and so on; and (2) the building of shared infrastructure through open architectures and test-beds that allow for grounded iterative experimentation in the context of real components, both human and technical. Such architectures could go a long way to increasing the feasibility and impact of experimental research in academia and to creating an ecosystem that supports iterative innovation.

Education and training within the target domains constitute an equally important goal. One challenge is in the translation of problems from one domain or field to another—for instance, describing the power and electric grid systems as a dynamical system and control problem—and then translating sometimes newly exposed assumptions back to the problem domain. Information and data are critical to understanding the challenges, formulating solutions, deploying solutions, communicating results, and facilitating learning and new behaviors that are based on results of the work. Thus a significant component of meeting virtually all sustainability challenges is to infuse computational thinking and computer science- and information-rich approaches into the deploying industry and the research and mission agencies.

PRINCIPLE: Undergraduate and graduate education in computer science should provide experience in working across disciplinary boundaries. Graduate training grants and postdoctoral fellowships should support training in multiple disciplines. Undergraduate and

---

[6]These shifts are already underway in various fields—for example, biocomputing.

graduate programs should include tracks that offer introductory and intermediate course work in such sustainability areas as life-cycle analysis, agriculture, ecology, natural resource management, economics, and urban planning.

Research institutions—both universities and the funding organizations—could better address the needs of authentic multidisciplinary research, in terms of publication venues, funding, criteria for promotion, research infrastructures that help enable sustained collaboration, and cross-training. The latter include the cross-training of students in multiple fields to enable them to bring a computer science perspective into other arenas. Authentic multidisciplinary work is challenging.[7] Work will need to be done across disciplinary boundaries and incorporating experts from many disciplines, as well as individuals with deep expertise themselves in more than one discipline. Examples of opportunities to enhance multidisciplinary approaches are described below:

- The creation of certificate programs, extension programs, and online programs for professionals in the target industries and agencies through professional societies and lifelong learning and training;
- Scholarships and fellowships both for computer science graduate students and for early-career professors that provide financial support for taking the time to develop expertise in a complementary discipline;
- The development of cross-agency initiatives (such as the collaboration of the National Science Foundation [NSF] with the Environmental Protection Agency[8]) that encourage interdisciplinary collaboration in relevant fields;
- Support for the development of new, cross-discipline structures (perhaps departments or institutes) between computer science and other fields that can create a new generation of students who are agile both in computer science and in fields relevant to sustainability;

---

[7]For a discussion of some of the challenges, see Sean Eddy's essay on "antedisciplinary" science (*Public Library of Science, Computational Biology* 1(1), doi: 10.1371/journal.pcbi.0010006), in which he notes: "Focusing on interdisciplinary teams instead of interdisciplinary people reinforces standard disciplinary boundaries rather than breaking them down. An interdisciplinary team is a committee in which members identify themselves as an expert in something else besides the actual scientific problem at hand, and abdicate responsibility for the majority of the work because it's not their field."

[8]An example is the joint National Science Foundation/Environmental Protection Agency establishment of two centers to study the environmental implications of nanotechnology, described in a 2008 press release: see http://www.nsf.gov/news/news_summ.jsp?cntn_id=112234.

- Institutional structures that support multidisciplinary and interdisciplinary teams focused on a problem or set of problems over an appropriately long period of time;[9]
- Internships and career paths and placement programs that encourage computer science students and postdoctoral researchers to work in relevant government agencies, non-governmental organizations, and industries;
- Coordination between academic research in computer science and non-traditional industrial partners—that is, beyond the large IT companies—to scope problems, help train students, and cross-fertilize ideas; and
- Regular, high-level summits involving computer science and sustainability experts—practitioners and researchers—to inform shared research design, assess progress, and identify gaps and opportunities.

The conceptualization of the bottom-up emergence of universality is relevant to researchers, university systems, and funding agencies in the following respects:

- *Researchers—emphasizing a bottom-up approach affects how researchers select and approach their problems and how they approach the training of their students.* Cross-training, learning other languages and vocabularies, immersion, and intensive and sustained collaboration are all important aspects of how research will need to be done. Although it may be essential to making real progress with respect to sustainability challenges, long-term commitment to a specific domain area is typically inherently risky for a CS researcher, because specific problems in sustainability may be addressed successfully but universal ideas or techniques may not necessarily materialize.
- *University systems—a focus on bottom-up approaches affects how universities incentivize and create the infrastructure for faculty to pursue sustained multidisciplinary efforts.* The computer science community has made progress in tenure and in the promotion of individuals who straddle disciplinary boundaries. Is such boundary crossing sufficiently encouraged and explicitly incentivized? Publication rates and pressures are higher than ever. Publication is essential to a successful R&D ecosystem, but does an emphasis on frequent publication have a negative effect on the pursuit of R&D which tackles difficult application domains that have not previously been processed and translated into computer science problems?

---

[9]One successful example of such an effort was the collaboration between computer scientist James Gray and astrophysicist Alexander Shalay on the multiyear effort (which, of course, involved many others) to develop the Sloan Digital Sky Survey (http://www.sdss.org/).

Moreover, much truly multidisciplinary work will require large teams and collaborations; is there appropriate recognition of CS contributions to large, multiauthor publications? Also, an evaluation of the productivity of junior faculty may need to extend to evaluating the impact of the researcher in the realm of sustainability in addition to the field of computer science. Related to promotion is the question of appointments—in what departments are multidisciplinary researchers appointed, and how can such appointments be handled so that the multidisciplinary nature of the researchers' work does not count against them in their home departments?

- *Funding agencies—emphasizing bottom-up approaches may affect how agencies structure multidisciplinary programs.* The National Science Foundation is a primary funder of research in computer science in the United States. The former Information Technology Research Programs at NSF and its current Cyber-enabled Discovery and Innovation Program have demonstrated the feasibility of programs with significant multidisciplinary aspects and the impacts that can result. But such programs provide for a minority of CS research, and in the committee's view, the sense of the community, as seen in review panels, program structures, review criteria, and so on, is not generally favorable toward funding domain-specific projects. One challenge is that typical reviewers of prospective research in computer science tend to want to see universality from the start, which presents a fundamental problem for the CS research community; funding agencies such as NSF can do little about this matter if the community does not adapt.

The committee is encouraged by the establishment of Science, Engineering, and Education for Sustainability (SEES) as an NSF-wide area of investment. SEES aims for a systems-based approach to "promote the research and education needed to address the challenges of creating a sustainable human future" and places an emphasis on interdisciplinary efforts. With its emphasis on interdisciplinarity and the involvement of NSF's Directorate for Computer and Information Science and Engineering, the SEES program offers an opportunity to demonstrate the depth of IT and CS innovation that the core discipline can offer and the rich and globally important problem space of sustainability.

There are ongoing opportunities for NSF to take advantage of the significant domain expertise in other agencies in order to pursue a strategy broader than programs that are crosscutting with other research directorates within NSF. Such a broadened strategy would involve programs that connect researchers with domain experts, practitioners, and projects in relevant mission agencies (such as the Department of Energy, Department of Transportation, Department of the Interior, and Department of Health and Human Services).[10] Furthermore, it is critical to ensure that funding structures support data sharing, encourage citability, and so on, but at the same time, an emphasis on sharing should be balanced with the need for researchers actually to collect data and to begin solving problems.

Another role that funding agencies can play is to fund longer-term projects and to be tolerant of risk, particularly in these multidisciplinary and cross-disciplinary, potentially high-impact research areas. It may be useful to target special funds that encourage the switching of focus to sustainability challenges and to incentivize grounded, domain-specific collaboration and training.[11]

> **PRINCIPLE: Refine funding and programmatic options to reinforce and provide incentives for the necessary boundary crossing and integration in CS research to address sustainability challenges. In particular, funding, promotion, and review and assessment (peer review) models should emphasize in-depth integration with data and deployments from the constituent domains.**

> **PRINCIPLE: There should be strong incentives at all stages of research for focusing on solving real problems whose solution can make a substantial contribution to sustainability challenges, along with in-depth metrics and evaluative criteria to assess progress.**

Another critical issue for structuring research is to build in evaluation tools for prioritizing efforts and evaluating meaningful impact. The committee offers an evaluative framework below.

EVALUATION, VIABILITY, AND IMPACT ANALYSIS

One of the greatest challenges in multidisciplinary research is to establish evaluation metrics that are both actionable and meaningful across the constituent disciplines. This chapter concludes by identifying methodological opportunities for optimizing research outcomes and impacts. Each of the recommended areas for evaluation necessarily incorporates

---

[10] An example of such a program is NSF's Industry/University Cooperative Research Centers Program.

[11] An illustrative example is the National Institutes of Health Mentored Quantitative Research Development Award (K25), which serves to fund quantitatively trained researchers so that they can learn about an area of biomedical science (see http://grants.nih.gov/grants/guide/pa-files/PA-06-087.html). These awards require the identification of a mentor in the substantive area and a plan for training; they provide funding for a commitment of at least 75 percent time.

interdisciplinary team members to assess accurately and to guide the potential for positive and significant impact.

One set of validation metrics would ensure that sustainability-oriented efforts related directly to key components of sustainability and were making significant progress on one or more of the legs of social, economic, and environmental impacts.[12] Specific metrics might include the following: metrics for analysis of environmental impact (life cycle, energy needed to create and execute a solution, as well as energy saved, and so on); metrics for analysis of economic impact (cost of implementation, resources used to create intervention, externalities); and metrics for equity and engagement across different stakeholder groups, taking into account their interests, values, and concerns, during both design and execution processes.

Many of these metrics involve humans, and so the assessment of research using such metrics will often involve techniques from the social sciences.[13] They also involve techniques drawn from other fields, such as life-cycle analysis from civil and environmental engineering. Any solution that claims to engage with and increase sustainability should be able to make an argument that addresses possible ways in which it may also negatively impact sustainability in any of its three overarching components (social, environmental, and economic). Moreover, proposals for projects that aim to make an impact should include a life-cycle analysis, and such analyses should be accounted for in the projects' budgets, as such analyses are a non-trivial effort to do well.

Scale analysis is critical to most information technology designs; it includes spatial scaling, temporal scaling, location scaling, and computational scaling. As demonstrated by the classic computer science talk

---

[12]As an example, although the challenges of sustainability are much broader than those related to climate change alone, in that domain a measurable impact is the goal of decreasing emissions so that carbon dioxide levels in the atmosphere are at or below a certain threshold. Being as specific as possible in terms of the anticipated impact on sustainability is critical. More broadly, one might consider the areas addressed in life-cycle analysis: global warming, stratospheric ozone depletion, acidification, eutrophication, smog, terrestrial toxicity, aquatic toxicity, human health, resource depletion, and land use. See U.S. Environmental Protection Agency, "Life Cycle Assessment: Principles and Practice," available at http://www.epa.gov/nrmrl/lcaccess/lca101.html.

[13]This is a characteristic of much of the evaluation done in human-computer interaction research and practice. Also very relevant here are design methodologies and approaches that seek to account for human values: for example, Value Sensitive Design, as discussed in B. Friedman, P.H. Kahn, Jr., and A. Borning, "Value Sensitive Design and Information Systems," in P. Zhang and D. Galletta (eds.), *Human-Computer Interaction in Management Information Systems: Foundations*, Armonk, N.Y.: M.E. Sharpe (2006), pp. 348-372, reprinted in K.E. Himma and H.T. Tavani (eds.), *The Handbook of Information and Computer Ethics*, Hoboken, N.J.: Wiley (2008), pp. 69-101.

question "But does it scale?," an early measure of the potential for the impact of a solution or system relates to scale. As sensors become more powerful (being able to measure many parameters at high frequency) and cheaper, a massive increase in the amount of data is expected. Each step or component of a proposed solution should be designed to work at scale. However, scale has new connotations with respect to sustainability that must be considered. In particular, many of the most problematic phenomena involved in sustainability challenges play out over timescales that are difficult for humans to comprehend, and many of the solutions make sense only when applied at scale. Here are some of the scalability questions that should be considered when evaluating a project:

- *Spatial scaling* requires articulating the minimal scale that can have impact (for example, touch points or density, geographic coverage, and so on).
- *Temporal scaling* incorporates real-time, human-time, and planning-time considerations; duration; timescale of relevance (onset, persistence, resilience); and the very challenging issue of addressing multiple-human lifespan timescales.[14]
- *Location scaling* addresses the applicability of the approach across a range of location-related contexts.
- *Computational scaling* refers to the tractability of the problem in terms of data generated and in terms of the anticipated footprint of the approach with respect to energy and dollars expended.

For each of these dimensions, the impact of a targeted innovation may well be difficult to quantify precisely. But it is the responsibility of the researcher at least to estimate and justify the anticipated, first-order measurable savings and efficiency improvements, or mitigated damages, from its realization. Is the proposed approach fundamental infrastructure? A game changer? The introduction of new computing technologies and concepts should be coupled with impact assessment (positive and negative) and follow-up study/assessment along with plans to integrate and iterate learning. Quantifying sustainability results as contributed by computational methods is a daunting challenge, especially given the current lack of data for real-world systems. Focused efforts toward creating publicly available data repositories that could be used to compare the effect of methods on the performance metrics chosen may prove useful in some domains. Without such data, there is little common ground for

---

[14]B. Friedman and L.P. Nathan, Multi-lifespan information system design: A research initiative for the HCI community, *Proceedings of the 2010 ACM Conference on Human Factors in Computing Systems*, New York: ACM Press (2010), pp. 2234-2246.

systematically comparing different methods and their potential benefits. It is particularly important for researchers to estimate outcomes and for the community to develop ways to assess impact, as these steps may have a dramatic impact on whether or not an approach can be appropriated and applied in other contexts.

CONCLUSION

Meeting the challenges of sustainability, as noted in Chapter 1, will require more than information technology, applications of clever technology, and computer science research. Indeed, at the heart of many global sustainability challenges are questions of resource consumption and standards of living. (See Box 3.3.) Nevertheless, the committee believes that

---

BOX 3.3
Toward an Information-Rich, Sustainable Future

Numerous analyses make clear that resource consumption is at the heart of many global sustainability challenges. At the same time, populations around the world are striving to improve their standard of living—and despite the efficiency improvements that also accompany development, that has inevitably meant increased resource consumption.

Efforts to improve efficiencies and substitute more sustainable for less sustainable materials and methods are what underlie much of the discussion in this report. However, there may be a broader sense in which information technologies and computational approaches can alleviate or mitigate the problem. Efforts to shift standard-of-living metrics from resource-intensive to information-intensive have the potential to be a significant lever in addressing global sustainability, although such shifts will increase the need for ever-"greener" information technology solutions themselves. In an increasingly information-rich and carbon-restricted world, finding ways to use information so that it both enhances perceptions and realities of standard of living and reduces resource consumption will be critical.

Examples of shifting to information-rich, less resource-intensive lifestyles include adjustments to transportation practices such as: information infrastructures that transform the convenience and trust of shared and alternative transportation modalities instead of private automobiles; improved technology in vehicles; transportation displacement such as telecommuting, social media, and e-commerce; and so on.

Although such examples emphasize opportunities to shift what counts as improvements in the standard of living for individuals, ultimately it is the policy choices and decisions, at local, regional, and federal levels, that will determine how many, if any, of these shifts are possible. Thus, organizational and governmental actions and decisions will have significant impact on whether a shift to information-intensive choices can happen in order to produce a shift in the way that society operates, to engender more sustainable outcomes.

CS and IT research has deep and fundamental contributions to make to these challenges. This chapter has argued for a bottom-up approach to research that values application-driven results while also supporting the iterative process that eventually leads to more universally useful contributions. The committee has argued for a series of validation metrics that explicitly explore the true impact of a piece of work in the arena of sustainability. Such validation metrics should include those that deal directly with humans, economics, and ecosystems and those metrics that engage with the concept of scale (a good first-order proxy for the universality that may not yet be present).

Information technology is at the heart of nearly every large-scale socioeconomic system—financial systems, manufacturing systems, energy systems, and so on. One important consequence, which has been the focus of this report, is that advances in IT have become critical enablers of change in these systems. The goal of this report has been to shine a spotlight on areas where information technology innovation and computer science research can help, and to urge the computer research community to bring its approaches and methodologies to bear on these pressing global challenges.

Appendixes

A

Summary of a Workshop on Innovation in Computing and Information Technology for Sustainability

INTRODUCTION

On May 26, 2010, the Committee on Computing Research for Environmental and Societal Sustainability held the Workshop on Innovation in Computing and Information Technology for Sustainability in Washington, D.C. The goal of the workshop was to survey sustainability challenges, current research initiatives, and results from previously held topical workshops and related industry and government development efforts in these areas. The workshop featured invited presentations and discussions that explored research themes and specific research opportunities that could advance sustainability objectives and also could result in advances in computer science (CS). Participants were also asked to consider research modalities, with a focus on applicable computational techniques and long-term research that might be supported by the National Science Foundation (NSF), with an emphasis on problem- or user-driven research.

This appendix summarizes the discussion of the workshop panelists and the attendees. The summaries of the four workshop sessions provided in this appendix are a digest both of the presentations and of the subsequent discussion, which included remarks offered by others in attendance. Although this summary was prepared by the committee on the basis of workshop presentations and discussions, it does not, in keeping with the guidelines of the National Research Council on the development of workshop summaries, necessarily reﬂect a consensus view of the committee.

The sessions at the workshop were entitled:

- Session 1: Expanding Science and Engineering with Novel CS/IT Methods: "The Need to Turn Numbers into Knowledge";
- Session 2: Understanding, Tracking, and Managing Uncertainty Throughout the Science-to-Policy Pipeline;
- Session 3: Creating Institutional and Personal Change with Humans in the Loop;
- Session 4: Overcoming Obstacles to Scientific Discovery and Translating Science to Practice.

The workshop agenda is provided at the end of Appendix A.

SESSION 1: EXPANDING SCIENCE AND ENGINEERING WITH NOVEL CS/IT METHODS: "THE NEED TO TURN NUMBERS INTO KNOWLEDGE"

Discussions during the first session of the workshop focused on the role of computer science in helping solve sustainability challenges. A broad definition of sustainability was employed. Vijay Modi, Columbia University, provided examples of sustainability areas where computer science could help address some challenges; Robert Pfahl, International Electronics Manufacturing Initiative, discussed changes in electronic systems and products to improve sustainability; Neo Martinez, Pacific Ecoinformatics and Computational Ecology Lab, explored the role of computer science in improving ecological sustainability; Adjo Amekudzi, Georgia Institute of Technology, examined planning and management issues around infrastructure; and Thomas Harmon, University of California, Merced, discussed water challenges.

Following are examples given of the ways in which computer science can play a role in addressing sustainability challenges:

- *Urban electricity consumption.* Gathering fine-grained accurate measurements and statistics on energy usage of individual buildings can be difficult, due in part to the variety and diversity of building types. With better measurements, one could develop a useful model of energy usage over the course of a day and find opportunities, for instance, to store extra energy throughout the day for use at peak times.
- *Infrastructure planning.* The planning and development of effective infrastructure are very difficult to do at scale for the time span required. Compounding these challenges is a dearth of data on how and where people actually live and what their movements are throughout the day.

This limited knowledge of the movement of people and the limited understanding of where infrastructure needs exist make it difficult to plan infrastructure accordingly. Advances in remote sensing, to improve understanding of the use of current infrastructure, can help cities and utilities to formulate better infrastructure planning.

- *Clean water.* Access to clean water is an ongoing and increasingly challenging problem worldwide. One of the more difficult components of this challenge is detecting water below the surface of Earth. Although detection of water at and just below the surface is well understood, technology for finding water at deeper levels is limited. Better sensing technologies are needed to help differentiate between sand, wet sand, water that is ooding the sand, and so on.

The examples above are a just a few of the areas in which computer science has contributions to make to sustainability. Workshop participants examined a wide array of sustainability challenges in which specific CS/information technology (IT) advances could contribute to resolving these challenges. In many cases, it is a matter of developing new approaches for turning raw data (numbers) into knowledge and, ultimately, prompting action that results in more sustainable outcomes. Research opportunities cited by workshop participants in the areas of ecological sustainability (that is, relating to diverse and productive biological systems), transportation, and water resources are described below, along with associated computer science challenges. The first session concluded with a brief examination of the policy challenges of interdisciplinary work and of turning knowledge into actionable items.

Electronic Systems and Products and Sustainability

The International Electronics Manufacturing Initiative (iNEMI) is a consortium of electronics manufacturers and affiliates focused on environmental issues in electronics.[1] Every 2 years, iNEMI creates a roadmap that charts future opportunities for and challenges to electronics manufacturing for reaching sustainability objectives.[2] The iNEMI efforts began by focusing on hazardous materials. The early goals of the consortium were aimed at eliminating chloro uorocarbons from the cleaning of electronics, removing lead from electronics, and reducing the use of halogenated ame retardants and polyvinyl chloride (PVC) materials. More recently, the focus has been on the complete energy use of products, as discussed

---

[1] A list of iNEMI members is available at http://www.inemi.org/news/council-members.
[2] The 2011 iNEMI roadmap is available at http://www.inemi.org/2011-inemi-roadmap.

below. Sound scientific methodologies are needed to take into account total trade-offs among con icting device requirements and to model long-term reliability and life of these devices.

Products that are recyclable, use non-hazardous materials, or minimize the use of energy and matter tend to be less harmful for the environment. Often there are trade-offs among these concerns. For example, using fewer hazardous materials may increase the resources needed to manufacture a certain type of equipment. When considering the size of items, there is often a trade-off of size for function. For example, cellular telephones have grown larger in recent years as functionality has increased. This matters especially with regard to calculating potential waste over a product's entire life cycle, although in the case of cell phones, the increased functionality may mean that other, even larger devices are no longer needed. Digitization is another example in which the functionality of electronics has decreased the amount of hardware needed. As digital music players have become more ubiquitous, compact disc players—and discs—are becoming less and less necessary.

Life-cycle analysis is key to understanding the complete energy use of products, including the energy used in mining raw materials, producing semiconductors and other components, assembly, transportation, and, ultimately, consumer use of the product. Computing research can assist in the tracking and understanding of all of these inputs throughout the life cycle of products.

"Green computing"—making computers themselves more environmentally friendly—plays a role in the reduction of energy consumption. For example, basic assumptions about computers' operating environments can be rethought, to yield significant energy savings. The 2011 iNEMI roadmap recommends that server farms and machines be redesigned so that the temperatures of server rooms can be increased in order to reduce the amount of energy required for cooling.

Participants noted that a holistic approach to technology is needed to contribute further to sustainability in electronics. Continued work in the following areas is needed: in digital semiconductor technology, work is needed in order to increase density and reduce cost; in the incorporation of sensor networks, work is needed to provide detailed energy-use data; in electronic packaging technology; and in innovation in CS and IT algorithms and applications. Additionally, participants suggested that standards may play an important role here.

Ecological Sustainability

Threats to ecological sustainability include loss of biodiversity, species extinction and invasion, and the exploitation of ecosystems. Each of these threats has consequences for the robustness, resilience, and stability of the respective ecosystems. Computer science can play an important role in enhancing the understanding of the consequences to ecosystems of particular courses of action by assisting in measuring the current impacts of actions and predicting future impacts on these ecosystems.

Databases play a crucial role in the understanding of ecosystems. For example, the Global Ex-vessel Fish Price Database of the Fisheries Economics Research Unit is a valuable, large data set.[3] The database provides information on hundreds of types of fish and their market prices over an extended period, thus enabling a better understanding of conditions in the oceans and of the potential effects of fishing.

Computing will also play a vital role in helping researchers and decision makers understand collected data, which come from a variety of sources. Hardware and software will be needed to help analyze large sets of heterogeneous data. Advances in modeling and simulation will also contribute to the understanding of the information collected. Ecological networks are complex, high-dimensional, non-linear systems. Therefore, simpler mathematical representations are not adequate. Ecological systems need to be simulated over time. Participants noted that currently, the various time series and relevant data for the simulation of an ecological system can only be summarized. More accessible data including quantitative information from simulations is needed so that others can use the data and contribute to the work. Some of the challenges created by large, heterogeneous data sets and researchers' resource limitations have been resolved with remote-computing capabilities (currently referred to as cloud computing). The shared resource of cloud computing can allow for simulations to be run much faster. Additional advances are needed so that data simulations can be stored easier and computing power can be more easily shared.

Interdisciplinary research on networks has led to a greater understanding of food webs and other ecological systems. For example, paleontological food web analysis has provided a better understanding of the network structures of current food webs.[4] Gaining an understanding of food chains on the globe over vast timescales can help provide researchers with a sense of how some kinds of ecosystems evolved. If economic

---

[3]The Global Ex-vessel Fish Price Database and its various uses are described in U. Rashid Sumaila, Dale Marsden, Reg Watson, and Daniel Pauly, *Global Ex-Vessel Fish Price Database: Construction, Spatial and Temporal Applications,* Fisheries Centre Working Paper #2005-01, Vancouver, B.C., Canada: University of British Columbia (2005).

[4]The Pacific Ecoinformatics and Computational Ecology Lab has done much of the work related to paleontological food web analysis. A list of its publications is available at http://www.foodwebs.org/.

information to account for things such as price and biomass can be incorporated into models based on the understanding of modern food webs, the effect of economic exploitation on ecological systems can be better understood. For example, in a simple three-species food chain (such as large fish eating small fish eating plants), adding economic information to the model also allows for a separation of the effects of exploitation by humans for economic reasons from the effects of human exploitation for the purpose of subsistence. Participants discussed how network-based analyses might be useful in other areas of sustainability. Rules derived about ecological networks, for instance, may also apply to energy and economic networks. Can useful comparison be made between economic and food networks? Does food function like money in any sort of actionable way?

Computing-enabled "citizen science" provides ways for volunteers to collect and report information from their own environments and to contribute to the sustainability of those environments. Citizen science programs have existed since the early 1900s, beginning with the Audubon Society's Christmas Bird Count.[5] Now, new mobile technologies and social networking tools make collecting and reporting much easier. Volunteers can easily collect data, for example, on a particular invasive species and send the information to experts to examine.

Transportation and Social Sustainability

Participants discussed the connections between traditional measures of sustainability, which may typically be functions of space and time, and measures of social sustainability. With social sustainability, as shown in Figure A.1, the sustainability footprint becomes the rate of change of quality of life as a function of one's impact on the environment. Participants argued that social sustainability can be and needs to be more rigorously accounted for in discussions about other forms of sustainability.

Social sustainability can be considered when looking at transportation sustainability, for instance. Definitions of "transportation sustainability" typically focus on moving items (people, goods, and information) in ways that reduce the impact on the environment, economy, and society. Transportation plans have been required in all major metropolitan areas since the 1960s. Although customer satisfaction has been included in transportation planning for some time, assumptions regarding customer needs are often incorrect. Traditionally satisfaction has been seen as a

---

[5]For information on and a history of the Audubon Christmas Bird Count, see http://birds.audubon.org/christmas-bird-count.

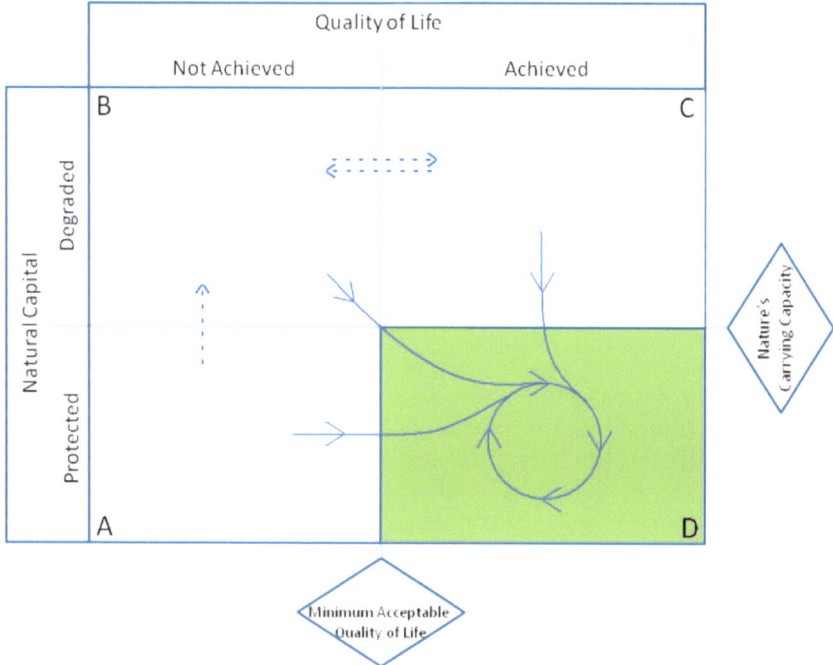

FIGURE A.1 Achieving quality of life within the means of nature. SOURCE: Jamie Montague Fisher and Adjo Amekudzi, Quality of life, sustainable civil infrastructure, and sustainable development: Strategically expanding choice, *Journal of Urban Planning and Development* 137(1):39-78 (2011). Reprinted with permission from the American Society of Engineers.

linear function of performance: for example, if a road is twice as smooth, customers are supposedly twice as happy. Research in this space has found, however, that this curve does not apply to all performance attributes. Gains in positive performance often have less of an impact on satisfaction, whereas reductions in negative performance are often more important to customers.

In this case, to build transportation plans that are sustainable both in the traditional sense and socially, customer satisfaction data, both subjective and objective, need to collected and woven into these plans. Data need to be collected on a wide range of attributes (safety, quality of life, smoothness), on the relative importance of the different factors, and on how customers rate the different attributes. Such information allows planners to distinguish between performance improvements that have positive and negative effects on quality of life and to negotiate the trade-offs between the two.

Computer science can contribute to such efforts by developing effective systems for collecting data from the public and providing better data-analysis tools to help, for instance, in the assessment of different choices regarding routes and other planning decisions. Real-time data processing and tools for planning and forecasting transportation needs can help urban planners and decision makers balance economic and policy challenges in planning future infrastructure.

Sustainable Sources of Water

Computer science research can help with the complicated problems of finding, tracking, and monitoring the sources of, need for, and sustainable use of water. Better sensors for measuring, better models for analyses, and better algorithms for optimization are all areas in which CS research can contribute. For example, more hydrological data and better models could help scientists to create a virtual watershed that would allow for quick studies of impacts and could potentially enable forecasts of the amount and quality of water available, much like weather forecasts.

In addition to creating virtual watersheds for analysis, areas in which improvements in CS and IT are needed in order to add to the understanding of water resources include the following:

- *Remote sensing.* Because it is not feasible to have sensors everywhere, models will continue to be important. Research is still needed on model-oriented science. Sensors, however, can be used to calibrate and fine-tune these models. A multiscale observation network can combine coarse-grained collection with more densely nested sensors deployed at a smaller scale.
- *Hyperspectral signal processing.* A wealth of information can be garnered from the re ected visible and non-visible energy from plants and water. Although much has already been learned from analyzing this information, more can be learned through a better understanding of the re ective spectrum patterns.
- *Spatial analyses.* Geographic Information Systems (GIS) and spatial analysis could be used for novel recognition and classification techniques and to identify the characteristics of an ecosystem. GIS imagery could also be used to detect shifts in an ecosystem.
- *Heterogeneous data integration.* Data combined from embedded sensors in rivers and from satellite images could provide a valuable picture of resources.
- *Workflows.* Tools are needed that are adept at scraping data from a variety of sources and combining them with spatial data repositories. The software tools could create input files and capture the history of simula-

tions so that researchers need not start from scratch. See the discussion in the section entitled "Scientific Workflows," below.

- *Computation.* Non-trivial optimization tools are needed in order to search for solutions to sustainability problems and to manage trade-offs. Powerful computing is needed to facilitate the scaling up of systems and to couple these with other contributing factors (economics, subsistence, and so on).

Policy Shifts

Although the solving of computing challenges will be one bridge to reaching further sustainability goals, challenges in interdisciplinary partnerships and in turning research into action and policy will need to be addressed as well. Participants noted that the traditional ways of building models tend to be incredibly time-consuming and isolating. Steps include the following: collecting the data, archiving the data, and selecting an individual (typically a doctoral student) to learn the model and then to deploy the model. A result of such an endeavor tends to be that several years later, only one person knows how to use the model effectively. Some progress might be achieved in this way, but to have a larger impact, computing support involving large data sets and complicated workflows will be needed. But such progress can only take place as far as unique partnerships across disciplines will push it. Participants noted that there tends to be a limited connection between the CS researchers and domain practitioners. More and better communication between the field and the laboratory could inform more useful research.

Better partnerships with computing and software experts could move research toward higher-impact results more quickly. As noted throughout the workshop, science and engineering are becoming increasingly dependent on software development. Fostering close collaboration between software experts and domain scientists is likely to be more effective than forcing domain scientists to learn advanced software engineering.

In addition to the computing research opportunities discussed above, participants urged that shifts be made in how research is translated into policy and action. More bridges need to be built between computer scientists and other disciplines, between researchers and practitioners, and between the academic and the industrial and the consumer settings. Technology from academic laboratories needs to move more quickly to the industrial and consumer world. This change would require collaboration and coordination at the research and development (R&D) level and the intervention of the research supply chain. With fewer and fewer industry-managed research labs, participants suggested that there has been a reduction in the integration of research and consumer products (of the sort that used to exist at Bell Labs) and that collaboration tends not to happen as smoothly. This lack of collaboration may prevent new technologies that would improve sustainability from reaching consumers.

Furthermore, planning and design are frequently done by economists, urban planners, and other decision makers, not by domain scientists or sustainability experts. Participants noted that these domain scientists need to be part of the process in order to provide feedback and more timely data, and they urged that academics more actively engage with policy makers.

SESSION 2: UNDERSTANDING, TRACKING, AND MANAGING UNCERTAINTY THROUGHOUT THE SCIENCE-TO-POLICY PIPELINE

When scientific information is provided to decision makers by the scientific community, explicit representation of uncertainty is rare. The loss of uncertainty information along the science-to-policy pipeline begins with the initial measurements, which may be recorded into databases just as numbers and without any additional information on how the data were captured or intercepted. From such a data set one might produce a predictive map, and any uncertainty that was captured may then be lost by means of an optimization process. Workshop participants noted that outputs from predictive and simulation models are often treated as exact or overly precise and accurate during policy making. In the end, without careful consideration of uncertainty, policy and decision mechanisms cannot be expected to achieve results.

The goal of the second session of the workshop was to explore some of the computational methods available to address loss of information about uncertainty, to consider what additional methods are needed, and to outline a potential research agenda. Panelists were asked to examine the following questions in relation to sustainability challenges during their talks:

• What are the sources of uncertainty that should be explicitly captured?
• What methods are suitable for explicitly representing uncertainty?
• Is the technological state of the art sufficient to model the many different avors of uncertainty present in large-scale sustainability problems? If not, what characterizes the types of uncertainty that are insufficiently modeled?
• What methods are suitable for assessing uncertainty in each stage of the pipeline? What shortcomings need to be addressed?

- Is the state of the art in human factors, interfaces, and computer-supported cooperative work sufficient to support the large-scale systems, models, and data sets that are necessary to tackle large-scale sustainability problems? If not, what needs are unmet?
- What are the appropriate techniques for working with uncertain data in data fusion, data assimilation, predictive modeling, simulation modeling, and policy optimization?
- How can explicit uncertainty representations be integrated into scientific workflow tools?
- Are there alternatives to explicit uncertainty representations that can improve the robustness of management policies to all of these sources of uncertainty?

Chris Forest, Pennsylvania State University, provided information on the sources of uncertainty and the tracking of uncertainty in climate models; Peter Bajcsy, National Institute of Standards and Technology, discussed the development of scientific workflows for tracking uncertainty through the science process; David Brown, Duke University, highlighted new methods for optimization problems under uncertainty; and John Doyle, California Institute of Technology, explored theories for analyzing "robust-yet-fragile" systems.

Assessing Uncertainty in Climate Models

Assessment and understanding of climate change and its impacts are critical to meeting many sustainability challenges. Scientists use a variety of techniques, including a variety of climate models, to assess and understand climate change. The potentially high impact of climate change means that policy makers are faced with hard choices, including but not limited to the reduction of emissions, adaption to climate change, and/or geoengineering that might help mitigate the effects of climate change.[6]

Participants discussed the role of uncertainty in the development and understanding of climate models. Scientists working on the problem of climate prediction must also address uncertainty. This could be done using a workflow plan that captures uncertainty information at each stage of the climate-prediction process. Within each stage, there are data, a model, predictions, assessment of likely impacts, and decision making. At each point there are sources of uncertainty that have to propagate

---

[6]National Research Council, *America's Climate Choices,* Washington, D.C.: The National Academies Press (2011).

through the system, ultimately leading to an estimate of the probability of the outcomes. Uncertainty analysis is driven by multiple goals, including the mitigation of climate change, adaption to changing environmental conditions stemming from climate change, and vulnerability assessments.

Assessing Uncertainty

There are two types of uncertainty in climate models: structural and parametric. (The level of uncertainties within each model creates a hierarchy of climate models, as described in Box A.1. Box A.2 then presents data summarized from the highly complex models used by the Intergovernmental Panel on Climate Change [IPCC].) *Structural uncertainty* stems from the hierarchy of models and attempts to balance the speed of the model with the complexity and components of individual models. Models take significant time to build; their complexity increases as more components are added. Modern tools and approaches in software systems, such as modularity, are important in creating current models. However, several of the models were built in the 1960s and 1970s before these tools existed. For example, participants observed that it is not possible to do comparisons between several of the older models, such as that of the United Kingdom's Met Office Hadley Centre and the National Center for

---

BOX A.1
Hierarchy of Climate Models

The first climate change assessments were done using the global energy balance model. Over the past 50 years, a number of additional types of climate models have been developed, creating a hierarchy of climate models. Each model typically has five major components—atmosphere, ocean, ice, land, ecosystems, and human action—and each component can be incorporated at various levels in the models, making the models significantly complex.

The most basic of the models is the energy balance model, which is very fast to run but lacks a lot of detail. In terms of complexity, the next level up includes Earth-system models of intermediate complexity (EMICs), which are reasonably fast and able to explore feedback and uncertainty. Fifteen years ago, the Massachusetts Institute of Technology created one of the first EMICs, which included all of the major components listed above. The model is very fast and can explore feedbacks between systems, is flexible enough to do uncertainty analysis, and can propagate uncertainty through the different stages.

At the next complexity level are the atmosphere-ocean general circulation models and Earth-system models, which are used by the Intergovernmental Panel on Climate Change (IPCC) (and from which the data for Box A.2 were drawn).

BOX A.2
Climate Change Observations and Climate Model Hindcasts

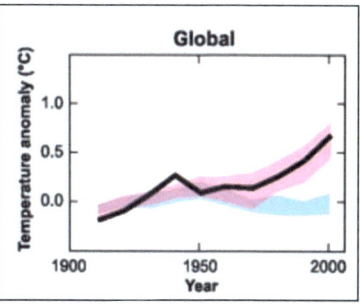
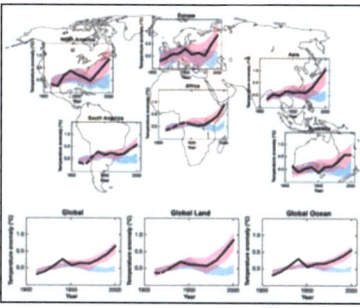

Figure A.2.1   Figure A.2.2

Figure A.2.1 shows a summary of the output from Intergovernmental Panel on Climate Change (IPCC)-class atmosphere-ocean general circulation models that were run for the Fourth Assessment Report of the IPCC. The widest bar represents the prediction of climate change of the 20th century, with the range of values representing 20 different climate models. The lowest bar represents predictions for the same models if anthropogenic climate forces are not included. The black line is observed temperatures. The comparison of these two lines provides the capability to assess the ability of the models to predict history.

Figure A.2.2 moves the models to the regional or continental level. The widest band, which represents the uncertainty in the predictions, widens. The bands still match the observational records, but this comparatively crude set of graphics typifies the extent to which uncertainty information tends to be portrayed to policy makers.

SOURCE: Intergovernmental Panel on Climate Change (IPCC). Summary for Policymakers, in *Climate Change 2007: The Physical Science Basis*. Contribution of Working Group I to the Fourth Assessment Report of the Intergovernmental Panel on Climate Change (S. Solomon, D. Qin, M. Manning, Z. Chen, M. Marquis, K.B. Averyt, M. Tignor, and H.L. Miller [eds.]), Cambridge, United Kingdom: Cambridge University Press (2007).

Atmospheric Research's community climate model, or the Geophysical Fluid Dynamics Laboratory climate models, by swapping in different components from each model; the software is too in exible.

Parametric uncertainty encompasses the adjustable parameters in a particular model. Model complexity and model expense limit the ability to do a full sampling of the parametric uncertainty space. There are numerous uncertainties in each model—those in observations, those stemming from natural variability in the climate system, and those in the model components themselves. As each of these parameters is added to the model, the model becomes less flexible. Various techniques have been tried for incorporating each of the uncertainties. However, these techniques have limited use because building adjunct models is as complicated as building a climate model itself.

Example: Integrated Global System Model

The Massachusetts Institute of Technology's (MIT's) Integrated Global System Model (IGSM) (Figure A.2), a coupling of a human systems model and Earth-system model, illustrates how uncertainty analysis is being applied to climate models.[7] The model uses several components: human activity; atmospheric, ocean, land, and ecosystem interactions; and biogeochemical exchanges. The model also runs comparatively fast. The following uncertainties are included in the IGSM: emissions uncertainty from MIT's Economic, Emissions, and Policy Cost model; climate system response (climate sensitivity,[8] rate of heat uptake by deep ocean, and radiative forcing); carbon cycle uncertainty; and trends in precipitation frequency. Climate sensitivity and ocean heat uptake, part of climate response, are a large source of uncertainty. Observational data can be used to calculate a probability distribution for climate sensitivity and the rate of ocean heat uptake. This calculation can be included in the MIT IGSM, and researchers can examine the resulting probability distribution of global average surface-temperature changes.

Explaining these uncertainties to decision makers is also a challenge that computer science may be able to help with. For example, researchers have compared the resulting global average surface temperatures with no greenhouse gas (GHG) policy intervention, or business-as-usual policies, and the resulting global average surface temperatures from implementation of GHG policy that limits carbon dioxide (CO_2) concentration to about 550 parts per million. To communicate the resulting difference in temperatures and the probability of the prediction, researchers created the

---

[7] A.P. Sokolov, C.A. Schlosser, S. Dutkiewicz, S. Paltsev, D.W. Kicklighter, H.D. Jacoby, R.G. Prinn, C.E. Forest, J.M. Reilly, C. Wang, B. Felzer, M.C. Sarofim, J. Scott, P.H. Stone, J.M. Melillo, and J. Cohen, *MIT Integrated Global System Model (IGSM) Version 2: Model Description and Baseline Evaluation*, Joint Program Report Series (July 2005), available at http://globalchange.mit.edu/research/publications/696.

[8] "Climate sensitivity" is the measure of how responsive the temperature of the climate system is to changes in radiative forcing; it is usually represented as the temperature change associated with a doubling of the concentration of carbon dioxide in the atmosphere.

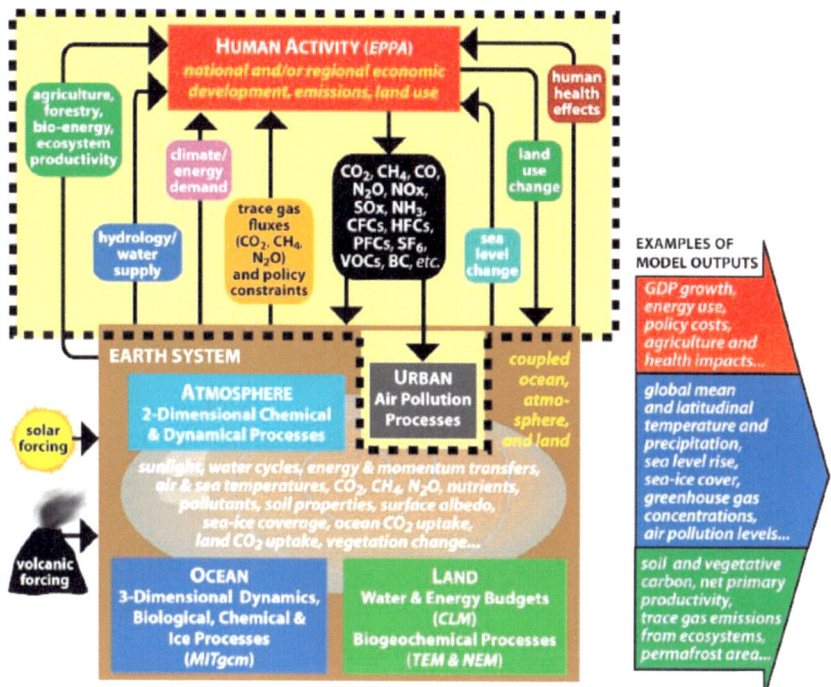

FIGURE A.2 Massachusetts Institute of Technology's (MIT's) Integrated Global System Model. NOTE: carbon dioxide (CO_2); methane (CH_4); carbon monoxide (CO); nitrous oxide (N_2O); nitrogen oxides (NO_X); sulfur oxides (SO_X); ammonia (NH_3); chlorofluorocarbon (CFCs); hydrofluorocarbons (HFCs); perfluorochemicals (PFCs); sulfur hexafluoride (SF_6); volatile organic compounds (VOCs); black carbon (BC). SOURCE: A.P. Sokolov, C.A. Schlosser, S. Dutkiewicz, S. Paltsev, D.W. Kicklighter, H.D. Jacoby, R.G. Prinn, C.E. Forest, J.M. Reilly, C. Wang, B. Felzer, M.C. Sarofim, J. Scott, P.H. Stone, J.M. Melillo, and J. Cohen, *MIT Integrated Global System Model (IGSM) Version 2: Model Description and Baseline Evaluation*, Joint Program Report Series (July 2005), available at http://globalchange.mit.edu/research/publications/696. Reprinted with permission.

Greenhouse Gamble roulette wheels (Figure A.3). This figure allows easy communication of prediction and uncertainty to decision makers.

Potential Contributions by Computer Scientists

Uncertainty analysis at the global scale is reasonably well understood, but increasingly there is a need to understand uncertainty at the regional and local scales. A better understanding of regional impacts of climate

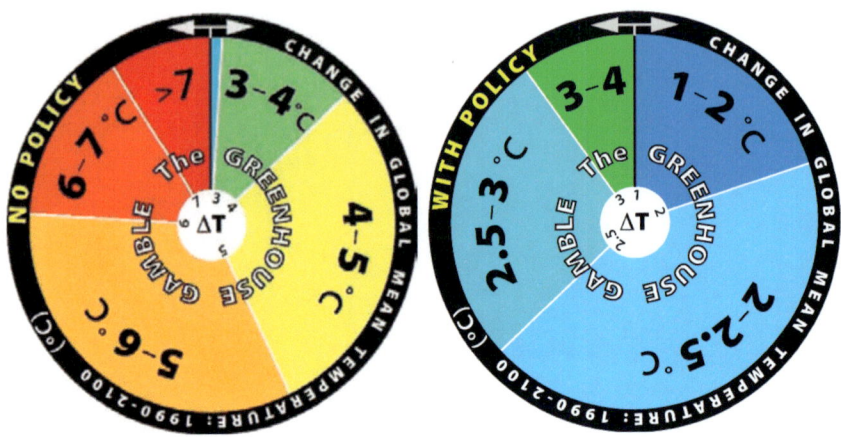

FIGURE A.3 The greenhouse gamble. Uncertainty can be represented by roulette wheels: (*left*) what could happen if no policies are adopted to lower greenhouse gas (GHG) emissions; (*right*) what might happen if GHG reduction policies are enacted. The size of each slice represents the probability that the coordinating temperature change will happen. SOURCE: Massachusetts Institute of Technology Joint Program on the Science and Policy of Global Change, available at http://globalchange.mit.edu/. Reprinted with permission.

change allows for better management of water resources, ecosystem changes, and air quality issues. The climate modeling community does not currently have the tools to sample the models for regional uncertainty information. Computer scientists are needed, for instance, to help determine what parameters in the climate system are driving uncertainty at the regional level, which are not the same parameters that drive uncertainty at the global level.

As noted in the section above entitled "Assessing Uncertainty," several models are quite old. Simulation code dates back to the 1960s and 1970s, much of it written in Fortran. This code needs to be redesigned to take advantage of advances in computing language, software modules, and interoperability.

Expert decision making is imperative to the climate modeling process. Although researchers seek to be as objective as possible in examining data and determining probability, the number of systems involved means that much calibration is done by hand. Experts must identify and rank uncertainties at each stage of the process, but experts have limited knowledge and will focus on what is known, while the edges and boundaries of the modeling system may be left unexplored.

Scientific Workflows

Participants discussed the typical nature of scientific workflows and the importance of uncertainty analysis to their effectiveness. Uncertainty information about data collected and generated for analysis is often unavailable. When it is available, there are no standardized data structures for sharing and managing this information. Due to the complexity of uncertainty modeling, there are very few software tools that can incorporate, compute, and propagate uncertainty information. If information regarding uncertainty can be propagated somehow, there are still challenges in visualizing and disseminating the uncertainty information.

One example of the difficulty of tracking uncertainty is in the use of multiple sensing and data-collection instruments. Often more than one type of instrument, such as a camera and a point sensor, is used in the same area or space to measure the same phenomenon. A question then arises: Which measurement is more accurate, where, and at what time? Using multiple instruments introduces several types of uncertainty, including that related to the transformation that researchers apply in order to display measurements, that related to spatial registration, and that related to the temporal synchronization. In order to grapple with these uncertainties, a theory of uncertainty is developed on the basis of a formalized framework that describes how to compute uncertainty. Researchers select a measurement based on the uncertainty level of each type of uncertainty. As the uncertainty framework is built and used, the error propagation rules and methodology allow one to build workflows that can be applied to future models and calculation.

A second example of the difficulty of tracking uncertainty is in managing and tracking data. Data often come from a variety of sources and are then processed by various computing techniques. As data move from analog instrumentation to a visual representation of research findings, uncertainty generally increases. For example, visualization features may be derived or elevation and/or slope may be computed. Moreover, there may be many users of the collected data, and each one may be using a different suite of software tools to process and analyze the data.

These examples provide some sense of the complexity involved in tracking and accounting for uncertainty. Some lessons become apparent. First, uncertainty information is vaguely defined; typically it is either a range or a distribution set. Second, the complexity of uncertainty modeling using error propagation is greater than that of the underlying phenomenon itself. Third, and most importantly, scientific workflows are suitable for managing uncertainty modeling; software modules could be reused and could track how collected data are being manipulated.

Defining Workflows

The traditional objective of scientific workflows is to automate a science process leading to a science product, usually a data set or a visualization. Workflows make it easier for scientists to manipulate, communicate, and reuse or repurpose data sets. Workflows also allow the computation to be done locally, or, when managing especially large or complicated data sets, scientists can take advantage of remote computational resources. As noted earlier, workflows can be used to track and manage uncertainty propagation and need to become part of the general scientific infrastructure. Capturing the workflow and managing the computation are particularly useful if all of the calculated information, including uncertainty information, can be delivered to third parties and end users.

Workflows can become a communication mechanism for the management of uncertainty. Using dynamic visualization and the sharing of workflows, scientists can more readily engage policy makers. Workflows can be designed with sharing in mind. For example, social media concepts such as tagging and networking can be incorporated into the designs of workflows, making them more accessible.

Research Questions and Challenges

Several open questions regarding the tracking and managing of uncertainty will have to be addressed before these uncertainties can be effectively calculated and, importantly, communicated using scientific workflow mechanisms. These include the following:

- How can uncertainty and information loss due to data translations best be captured?
- How can provenance information about uncertainty methods and parameters best be gathered automatically during computations?
- How can uncertainty best be propagated in computation workflows, or perhaps, how can uncertainty propagation rules best be offered for software tools without such rules?
- How can uncertainty best be delivered and presented to decision makers, who may require a customizable view so as to increase its effectiveness, as well as a universal viewer for other interested parties, who may require more widely accessible information?
- Can workflow services managing the uncertainty be integrated with other web services, such as mapping or sharing tools, to deliver uncertainty to scientists and policy makers?

Participants argued that a CS research agenda for sustainability needs to support research on the representations and propagation of uncertainty. In the past there has been little information on uncertainty. Content managers should be encouraged to include uncertainty with data, use workflows to manage and track uncertainty, and incorporate other information-sharing services, ultimately leading to better information on which decisions can be made.

Robust Optimization Under Uncertainty

Optimization has progressed rapidly in recent decades due in part to the rapid improvement of computer systems generally and to the development of increasingly sophisticated algorithms. The solving of large-scale, linear problems, with millions of variables, is computationally feasible. However, the inclusion of and calculations regarding the uncertainty in these optimization problems is still a challenge. For policy decisions, for instance, models and calculations are run many times, and changes are made during each iteration. Because uncertainty has to be calculated at each iteration, running models that include uncertainty is extremely inefficient. The goal is to make accounting for uncertainty computationally tractable so that each model can be run faster and more efficiently.

Robust optimization is one method for coping with uncertainty in optimization problems. Robust optimization provides computational tractability and supports parsimonious modeling demands—one does not have to worry about the specifics of probability distribution. Robustness is an inherent and essential feature of many important methods across many disciplines, including machine learning and decision theory.

Robust optimization is different from sensitivity analysis. Although robust optimization and sensitivity analysis are motivated by similar factors, sensitivity analysis is a post-optimization tool; if robustness is ensured beforehand, solutions will not be overly sensitive. Robust optimization is also sometimes considered too conservative. The conservativeness relates to the uncertainty set that is used and on how large it is. An improved data-driven theory of optimization is needed. There are many approaches to building uncertainty sets, but there is the open theoretical question regarding the right way to use data in optimization problems. In some problems, when there are not enough data, the questions become these: How does one properly incorporate subjective opinion about the data? What is the right way to describe uncertainty? An additional challenge is that idealized problems tend to be studied without enough application to real-world problems. There is also the challenge of ensuring that uncertainty is acknowledged and taken into account in any decision-making process.

Theory and Methodology of Robust-Yet-Fragile Systems Analysis

Participants noted that fundamental research is needed to improve the understanding of the various trade-offs in computational methodologies, such as efficiency versus robustness. Complexities of real systems—not the complexity of the mechanisms used to study the systems—embody these trade-offs. Some theories already exist for examining the trade-offs among robustness, fragility, and efficiency. For example, researchers know that efficiency has hard limits and is bounded. Robustness and fragility have conservation laws as well, and the trade-offs between the two are tangible. Theories from other disciplines, including systems engineering, control theory, information theory, and computational complexity theory, provide complementary approaches and can be integrated into theories on robustness, fragility, and efficiency.

Efficiency and robustness exist in many dimensions, and although each in itself is reasonably well understood, a theoretical framework is needed for conveying the interactions and examining the inherent trade-offs. Firm trade-offs exist among the following:

- Efficient use of resources (sustainability)
  —Small amounts of resources consumed, small amounts of waste produced.
  —Inexpensive components, small capital investment.
  —Efficient processes: design, manufacture, maintenance, management.
- Robustness to perturbations
  —Rejection of external disturbance and suppression of internal noise.
  —Tolerance for component failures and uncertainty.
  —Security against malicious attack and hijacking.
  —Scalability to large system size.
  —Evolvability on long timescales to large changes.
  —Human actors with aligned incentives.
- Predictable, verifiable, understandable
  —Limits on unintended consequences.
  —Easily reproducible experiments and data.
  —Models (simple and analyzable), short theorems, proofs.
  —Experience that is a reliable guide to the future.[9]

One can start building toward a theory with comparisons across disciplines. Although efficiency limits are understood, it is very difficult, if not impossible, to reach 100 percent efficiency rates. By contrast, robust fragility is much less understood. Robustness in one part of a system may induce fragility in another. Fortunately, evolvability and robustness seem

---

[9]John Doyle, "Theory and Methodology of Robust-Yet-Fragile Systems Analysis," presentation at the Workshop on Innovation in Computing and Information Technology for Sustainability, Washington, D.C., May 26, 2010.

to be compatible. Architectures and platforms that enable innovation well can also enable robustness well.

Examples were discussed of how new theories can be developed to encompass needed robustness requirements. Network theory might be able to help sort out what is an accident of similarity and what is deeply structural. However, the fact that reasonably well understood networks exist in one domain does not mean that these understandings translate well to another domain. There may be useful knowledge to be gained by comparing network structures and properties in different domains (for example, contrasting and comparing climate history, cell systems, and Internet architectures), but this knowledge needs to be validated first. Additionally, participants noted that a better understanding of what is meant by complexity, non-linearity, modularity, architecture, and evolvability in different domains is needed so that scientists can communicate more effectively with policy makers and with one another.

Furthermore, big data and big models mean that many things can be demonstrated by means of data or models. As computational capacity has increased, in many ways research efforts have moved from coping with impoverished data and elegant models that are not well understood to coping with massive data sets and sophisticated simulations that are not well understood. In this new environment, larger gaps between the demonstration and the reality can be unintentionally created, and unanticipated fragilities may become overwhelming.

SESSION 3: CREATING INSTITUTIONAL AND PERSONAL CHANGE WITH HUMANS IN THE LOOP

Behavioral changes at both the institutional and the individual level are needed in order to achieve sustainability objectives. Important questions in designing and developing smarter systems involve the level of information and the interface design that will induce behavior change. Human-system interaction (HSI) issues arise both for individuals in homes and offices and for administrators of larger systems and facilities. These interactions can occur at different timescales, encompassing both day-to-day decisions made by users and operators and planning decisions involving longer periods of time. Moreover, although there have been many advances in HSI, the literature is replete with failed cognitive models, serving as cautionary tales for HSI in sustainability applications.

Panelists were asked to examine the following questions in relation to sustainability challenges during their talks:

- How can data and information be presented at the appropriate granularity and time frame to be most effective? What system, application, and user factors bear on the human-system interaction design choices?
- Describe the potential impacts of the various contexts and trade-off decisions that might need to be made, including the following: the impact of context (e.g., stakeholders, and so on), the impact of large versus small groups versus individuals, the impact of income, the impact of use by or for cities versus businesses versus individuals, the role of middleware, the supply chain, and so on.
- How do human factors affecting sustainability challenges drive the use and design of technology? How can this interaction be accounted for? When are power, networking, products, and other information and communication technologies (ICTs) really needed? Discuss human choice and its impact on consumption, disposal, and reuse.

Bill Tomlinson, University of California, Irvine, examined current use and research on computing initiatives that provide information for more sustainable decision making; Shwetak Patel, University of Washington, explored the challenges of providing utility-use data to consumers; and Eli Blevis, Indiana University, examined the possibility of incorporating sustainability ideas into the design process.

Better Information for More Sustainable Decision Making

One way to use information technology to lead to more sustainable outcomes is through the provision of information to individuals and organizations to assist in decision making. Participants noted that this type of assistance can be provided at many levels. They discussed various types of tools ranging from those that can help inform comparatively simple sorts of personal decision making; to tools that can help people understand large, complex challenges and how their individual actions and behaviors might help; to tools that can assist in understanding and resolving complex challenges that require complex and/or coordinated responses.

Several examples of how IT can be used to perform small-scale sustainability-related tasks more effectively were described. They include the following:

- *Cellular telephone use for coordinating the sale of fish catches.* During the early 1990s, fishermen in Kerala, India, would typically have 6 to 8 percent waste from their catch. The waste was primarily due to lack of buyers for these fish. Cell phone availability in the late 1990s quickly changed this mismatch. As they were coming ashore, fishermen could more easily locate buyers, and waste was quickly eliminated. Cell phones were not developed to tackle environmental issues, but did enable this unintended positive outcome.

• *The use of various communication tools by non-profit and non-governmental organizations in coordinating activities.* The Surfrider Foundation[10] organizes beach cleanups using various IT infrastructures, including text messages, e-mail campaigns, and social media tools.

• *Consumer tools for a smart grid.* Researchers are examining ways to apply Internet-inspired architectural principles to the energy grid. For consumers, questions are being explored regarding smart appliances that automatically run when electricity is in less demand or when renewable sources are available.

Another class of tools is designed to help educate the public on their choices regarding sustainability. Examples discussed by participants range from tools for educating school-age children about ecological interdependency, to those for helping consumers make more sustainable purchases at the market. The Social Code Group, which Bill Tomlinson leads, has developed several of these tools. A selection of these tools and projects was discussed:

• *EcoRaft.*[11] This application was designed, with contributions from ecologists, to help 8- to 12-year-old children learn about ecology in the museum environment. A large monitor and several tablets represent various ecosystems. The tablets allow interaction and can be used to simulate the transplanting of species from one ecosystem to another, encouraging children to explore the interdependencies among species and the interconnections between restoration and conservation. A key aspect of EcoRaft is a button at the bottom of the main screen that would remove all species from the simulation. In science museums, the first thing that children tend to do with interactive displays is to push whatever buttons are available. In this case, the button was not labeled, which meant that current users had to guard the button or educate newcomers to the game, simulating the value of education and activism.

---

[10]Information about this organization is available at http://www.surfrider.org/.
[11]Bill Tomlinson et al., The EcoRaft project: A multi-device interactive graphical exhibit for learning about restoration ecology, in CHI'06, *Extended Abstract on Human Factors in Computing Systems,* New York, N.Y.: Association for Computing Machinery (2006). DOI: 10.1145/1125451.1125717.

- *GreenScanner*.[12] This tool was developed to help consumers when making purchasing decisions to understand the environmental impact from the growing, processing, and transporting of their foodstuffs. The vision was that using cell phone cameras and large databases that linked universal product codes (UPCs) to environmental data, consumers could quickly identify which available foods had the least environmental impact. When GreenScanner was first developed as a web application in 2006, sufficiently capable hardware was not in wide use to make this tool feasible for everyday consumer use. However, as cell phones have advanced, a tool similar to GreenScanner has become commercially available through the company Good Guide.[13]
- *Trackulous*.[14] This tool was designed to help people track their activities and the environmental impact of those activities. People may be aware that they often travel by airplane or car, but they might not understand the cumulative time spent doing these activities in a year or the cumulative impact that such travel can have. By tracking their activities, people can better understand their carbon footprint and where the opportunities are for lessening that footprint.
- *Better Carbon*.[15] This web application uses collaborative filtering techniques to reduce the amount of information that a user needs to input into carbon-footprint calculators. With current carbon-footprint calculators, users have to enter a great deal of information before receiving an answer. With collaborative filtering, they can enter much less information (which is compared to similar information provided by other users) and then be provided with meaningful defaults for the additional information required.[16]

Participants noted that tools such as these can help individuals better understand their contributions to sustainability problems as well as to sustainability solutions. However, a complete toolbox for resolving large, complex environmental problems does not yet exist. The scale of environmental challenges on the planet today—including global climate change, biodiversity loss, sea-level rise, and various kinds of pollution—is much greater than the scale of other challenges that humans typically face. People are generally good at cooperation in small-scale tasks and at understanding and resolving smaller challenges, especially those with

---

[12]The web application, GreenScanner, is available at http://www.greenscanner.net/.
[13]The tool is available at http://www.goodguide.com/.
[14]The tool is available at http://trackulous.com/.
[15]The tool is available at http://www.bettercarbon.com/.
[16]Collaborative filtering systems are used in other contexts; for example, Amazon uses such tools to recommend products based on past purchases, and Netflix uses them to recommend movies that users might like.

smaller scales of time, space, and complexity. However, the largest and most challenging environmental problems are not at scales that people can readily understand. Examples of the complex scales involved in sustainability include the following:

- *The timescale of a rise in sea level.* Consider the prospect of the sea level rising, perhaps 40 centimeters, over the course of 120 years. How can the general population understand the risks and consequences?
- *The geographical scale of the supply chain.* Another example of a scale that is difficult to comprehend is global supply chains. Products purchased in the United States arrive through a supply chain that may begin as far away as a palm oil plantation in Sumatra. How does one communicate to the consumer the potential environmental repercussions of the manufacture, transport, and life cycle of products that they purchase?
- *Climate change.* The complexity of climate change stems from the large number of factors—such as cloud cover, carbon dioxide emissions, rainforest depletion—and the interaction of these factors that make up the global climate system. How can IT—and the way that sustainability information is communicated using IT—assist in developing an understanding of the increasing scales of time, space, and complexity that characterize climate change?[17]

There may be lessons to learn from general IT and CS approaches when it comes to these complex problems. Good architecture design can remove extraneous decision-making requirements and afford consumers the ability to select sustainable options quickly and easily. When the Internet works well, for example, users do not typically know or need to know where the information they are receiving is coming from: e-mails move through various wires, routers, and servers around the globe; web pages are populated by widgets whose data could be sourced from anywhere. One attendee suggested that analogous architectural approaches may be applicable to certain sustainability challenges.

Another big challenge in sustainability issues is the legacy and fixed nature of our utilities and infrastructure. There is a great deal of leg-

---

[17] There are several challenges in educating the public on climate and sustainability issues, one of which involves differentiating between weather and climate. It is too easy to overinterpret extremely hot or cold days. Interactive design techniques could help communicate the difference as well as the subtle scale changes, such as highlighting the shifting state of Alaskan glaciers or the sea level. The climate research community also struggles to define what is "normal" for weather and climate: "normal" from 1950 to 1980 is different from what would be considered normal from 1980 to 2010. Some advocate taking a first-order linear trend of the past 30 years as an estimate of a baseline to use to compare changes.

acy infrastructure in most systems that pose significant challenges. For example, one can certainly contemplate autonomous vehicles, which are much more environmentally friendly, but moving from a highway of all individually controlled vehicles to a highway of all autonomous vehicles is difficult; the upgrade path is bumpy, to say the least.

Residential Energy Measurement and Disaggregated Data

Participants discussed how better information can assist consumers in making effective changes in their use of home resources such as power, water, and gas. Current literature suggests that high-granularity or high-resolution data—for example, information about the usage patterns of individual appliances—are the most useful. Participants observed that literature from the past several decades suggests that if this information were provided to consumers, a 15 to 20 percent reduction in consumption would be possible.

In the past, the technologies that collected and provided this information were typically tedious to install and required installation by trained technicians. Such difficulties make them impractical for large-scale deployment. Tools were not successful in the past because too much burden was placed on the individuals who were installing them in their homes. New technologies, such as embedded systems and sensors, can make information gathering and feedback tools much more practical. An ongoing research challenge is to determine what information and interfaces work best. One cannot start answering more in-depth questions about the effectiveness of novel, engaging, and persuasive feedback applications, however, until there are more information and data from end users. Sensing and feedback work hand in hand in creating the most effective tools for consumers.

Challenges of Collecting Usage Data

Consumers, appliance manufacturers, and utility companies can all benefit from any data collected from in-home deployment of sensors. Consumers could use disaggregated data to understand their utility use and to make changes in it. By contrast, what is typically available to consumers is only simple end-of-month billing, which provides very little actionable information. If consumers knew the consumption of resources by individual devices, they might start to tailor their behavior for more efficient consumption. Use data could also help manufacturers gain a better understanding of the use of their appliances. Manufacturers do not have a lot of information on how often their appliances are used. Better understanding of typical duty cycles could be incorporated into future designs.

To the extent that utilities are motivated to promote energy-conservation activities, they have few ways to assess or validate whether those activities are working. Relevant data for utilities would include device use over time and any regional difference in use. Currently utilities tend to use self-reporting and polling to determine whether or not their initiatives in uenced consumer activities. Another challenge is that utilities have little experience in deploying technologies inside the home. Moreover, utilities do not want the added expense of installing monitoring tools in the home. Sensors that are easily deployable by end users could provide validation and verification of usage for utilities as well as for consumers.

Once validation and verification are widely available, utilities would be able to provide better incentives to customers for conservation. Additionally, the information could allow utilities to create better demand-response models. Better usage models could also be developed by researchers. One goal would be to create a national energy data corpus, which would be very useful to researchers across disciplines as well as in helping meet large-scale energy and sustainability information needs.

The research discussed in this session of the workshop focused on the creation of technology that (1) provides highly granular, disaggregated data on home energy use and (2) is deployable by end users. The traditional way of collecting such data would be to deploy a network of sensors at each outlet, light, or water fixture—a method that is cumbersome, expensive, and, as past experience shows, something that consumers are unlikely to do. This suggests the need to find an approach that is easier from the perspective of the consumer and, ideally, is a device that plugs in to a single outlet yet monitors electrical consumption or events that occur in the home at the appliance level.[18] Below are examples of research being done along these lines to collect data on three main resources: power, water, and gas.

- Research is being done to examine the propagation of electrical noise over power lines to infer what devices are being activated. This research incorporates work in signal processing, machine learning, and embedded systems. These types of devices are better than smart meters, which may only provide information every 15 or 20 minutes. In certain circumstances, homeowners or researchers may want shorter time inter-

---

[18]Tools built directly in to appliances are not particularly helpful because the timescale for replacing many in-home appliances is often measured in decades.

vals. To provide real-time consumption, a magnetoresistive sensor was developed that attaches outside the breaker panel, but can be read inside the home. This tool has been field-tested with some success by utilities to determine whether consumers are capable of installing it.

- Measuring and tracking water usage provides similar challenges. Obviously, any sensor whose installation requires cutting into pipe will not work for consumers. A single-point water-sensing solution, attached to a washing-machine hose spigot, could look at the water hammer phenomenon that occurs when a valve is opened and closed, which could infer which water fixture was being used inside the house. Because the hot- and cold-water systems are interconnected, one could even begin to discern hot- and cold-water use with a single sensor. In apartment complexes, individual units may not have hose spigots. In that case, the hot-water heater would be another location for installing a sensor.
- Collecting gas data is much more difficult. There is not typically a good place to locate a single sensor that determines when an appliance is activated, and there are safety concerns about customer-installation of gas fixtures. However, a sensor could be installed by clipping it to the nationally mandated regulator on appliances. The acoustic vibrations of the diaphragm are linearly proportionate to ow. By measuring these vibrations, real-time consumption information can be determined. In addition, transient events that occur from the opening and closing of gas valves manifest themselves through the valve itself. This information can be used to determine exactly what appliance is being used.

Interfaces for Actionable Feedback

Participants noted that interfaces for providing the disaggregated sorts of data and feedback discussed above are critical to the successful and effective deployment of sensors. The typical "interface" is the simple electric bill and meter in homes. The meter was not designed for consumers to read, and the bill, which *is* for the consumer, usually provides only an end-of-cycle reading. Even the units used in the typical electric bill are not very user-friendly: do most consumers understand, for example, what a kilowatt-hour means? Water bills are typically even more aggregated; most include up to 3 months of water use. If a spike occurred in water usage due to, for instance, a leaky valve, the consumer would not have any knowledge of it until the end of the billing cycle. Clearly there are opportunities for interface design to play a role in communicating better and more actionable information. Unfortunately, there are currently only limited examples of interfaces that provide disaggregated feedback in approximately real time. One obvious example is gas-electric hybrid automobiles, which display approximate instantaneous mileage while the car is running, as well as relative use of engine and battery. Given more information about their own driving habits in accessible and understandable fashion, consumers can (and do) alter their behavior and conserve gas, participants observed.

Participants described how research that validates the functionality of the devices described earlier has started with in-home trials. In order to evaluate the installation process, researchers have observed consumers self-installing devices. Once the devices are in place, researchers can evaluate the effectiveness of the energy-management systems provided to users. Some of the questions being asked by researchers are: Can you elicit behavior change, up to 15 to 20 percent reduction in usage, with disaggregated feedback, and does the behavior change really hold over time? As more data have been collected on how users react to various interfaces, researchers have begun to iteratively refine individual interfaces. Some of the deployed devices were also being integrated with Microsoft Hohm and Google's PowerMeter projects.[19]

Before devices and systems such as those exemplified here can be deployed commercially, large-scale studies are needed. Utilities are open to participating in large-scale studies, but when deploying research technology it can be a challenge to create enough devices to do even small-scale deployments in 20 or 30 homes; research laboratories are not often equipped for production. One solution has been to commercialize the technologies being used in research settings so that devices are available from retail stores as well as directly from the utilities. For example, Belkin International, which sells a wide array of computer peripherals and other consumer electronics, recently purchased a demand-side monitoring solution developed at a university research laboratory.[20]

The timescales and complexity of general sustainability issues are difficult to understand. Feedback mechanisms are being developed, and there is some evidence that information does change behavior. Additional knowledge is still needed on how much this feedback will in uence behaviors and what the best feedback over what timescale produces the most change.

---

[19]In June 2011, both Microsoft and Google announced the discontinuation of these projects. Both cited the slow market adoption of the services as the reason for the termination.

[20]For additional information, see http://seattletimes.nwsource.com/html/technologybrier dudleysblog/2011667981_uw_gets_slice_of_profs_startup.html. Examples of the Belkin products can be seen at http://www.belkin.com/conserve/.

Sustainable Interaction Design

Participants also discussed some general high-level principles regarding sustainable interaction design—the notion that the design of systems should incorporate sustainability considerations. A few design principles for sustainable interaction design, including the following, were discussed:

- *The connecting of invention and disposal.* Any new design should also include information on what will happen to the materials or products that it replaces.
- *Encouragement of renewal and reuse.* Human-computer interaction (HCI) research can play a large role by highlighting the future value of objects.
- *Encouragement of quality and equality.* The second and third user of any product should receive the same satisfaction as the original owner.[21]

A number of projects incorporate at least some aspects of sustainable interaction design, and several are described in the previous sections. Additional examples include corporate research done by SAP, Inc., and academic work done by HCI researchers at Carnegie Mellon University (CMU). SAP has developed Sourcemap,[22] which provides a way to track and improve supply chains. Using an open-data platform, users can track where each of their foodstuffs comes from. For example, a catering company can provide a map showing the route by which of all of its products were shipped, or individuals can determine how far each of their breakfast items traveled. Stepgreen,[23] created by CMU's HCI researchers, is another example that allows users to track the benefits of making sustainable choices. Users can identify sustainable actions that they are already taking, such as turning off lights not being used or walking to locations less than a mile away, or they can commit to future sustainable choices. Stepgreen tracks the amount of savings in dollars and carbon dioxide emissions for each action.

Interaction design can also be used to encourage dialogues among scientists, decision makers, and citizens. Interaction designers can help bridge the gap between scientific knowledge and public perception, can build support, and can promote discourse leading toward solutions. Par-

---

[21]Eli Blevis, Sustainable interaction design: Invention and disposal, renewal, and reuse, in *Proceedings of the SIGCHI Conference on Human Factors in Computing*, New York, N.Y.: Association for Computing Machinery (2007).

[22]See http://www.sourcemap.org/.

[23]See http://www.stepgreen.org/.

ticipants noted that bridging this gap has been a particular challenge in climate science and that there is an opportunity for interaction design to play a role. For instance, whereas some of the previously mentioned tools can help people make more informed choices about reducing their negative impact on the climate, tools are also needed that inform people's preparation and responses to climate change as a chronic sustainability challenge. Interaction systems are also needed to deal with the likely increase in the severity of natural disasters and crises due to climate change.[24] Participants discussed the concept of a "Dashboard Earth" that could be used to provide information about what is happening and where. Interaction design can also be used to develop interactive systems that could help with orderly evacuation in a natural disaster, for example, providing the information to local, regional, national, and intergovernmental policy makers about who can go where and how many people each location can absorb. Interaction design can be used to persuade and show people how to live with fewer resources, as matters of sustainability and preparation and adaptation. The interaction design of tools like social media is needed to help persuade people and various levels of organizations to care for others in the face of climate change and its effects. Interaction design can also assist in the public discourse about and support for certain sorts of solutions and in fostering public understanding. Participants argued that all of these tools and more need to provide information at the local, regional, national, and global level and to help people respond efficiently during crises.

SESSION 4: OVERCOMING OBSTACLES TO SCIENTIFIC DISCOVERY AND TRANSLATING SCIENCE TO PRACTICE

Committee member David Culler, University of California, Berkeley, and David Douglas, National Ecological Observatory Network, led the discussion during the final session, which highlighted some of the impediments to developing and deploying innovative information technologies for sustainability challenges. Guiding questions for this session were as follows:

• What are the motivations for and impediments to applying innovative information technologies to sustainability challenges and how do they differ by domain?

---

[24]National Research Council, *Adapting to the Impacts of Climate Change*, Washington, D.C.: The National Academies Press (2010).

- How can large-scale science addressing real-world problems be made credible, if reproducibility is not possible?
- What lessons can be applied from the transformation of the Internet into a critical infrastructure that must avoid ossification?
- What is the appropriate mix of empiricism, innovation, and application in order for computer science to have an impact in the area of environmental sustainability?

The Energy Challenge

Participants suggested considering broad sustainability challenges in the context of the energy challenge. The interconnected nature of people's basic resource needs, such as water, energy, and transportation, and the economic arrangements among these resources create a very complex problem. However, these interactions also mean that the energy challenge can serve as a useful proxy for sustainability challenges related to other limited resources.

The primary function of the electric grid is to deliver high-quality, low-cost power to millions of customers who are geographically distributed over thousands of miles. The fact that consumers have been able to make use of the grid without needing much knowledge about their own consumption patterns, or about where the power is coming from, has contributed to rapid economic and industrial growth. People have been able to use a comparatively inexpensive resource—energy created mostly through the burning of fossil fuels—essentially indiscriminately to expand the production of products that spur the economy. Additionally, enabling a usage model in which consumers could remain ignorant of their own consumption patterns meant that the grid has been tasked with delivering a high-quality commodity at extremely low cost. Moreover, the expectation has been that power would be delivered immediately as needed. The power grid is expected to meet these goals with minimal forecasting or anticipation of that need, except at very coarse granularity, and without inventory storage along the energy supply chain.

The current energy model is increasingly complex, with numerous sources of energy, a variety of stakeholders and consumers, and a not insignificant fraction lost during transport. A pressing sustainability challenge revolves around these questions: How can energy use be reduced, and can it be done without significant economic hardship? The following question was discussed: Where and how can computer science fit into this picture?

Figure A.4 shows the percentage of energy use in the United States by type. Each of these types represents an opportunity for reduction in

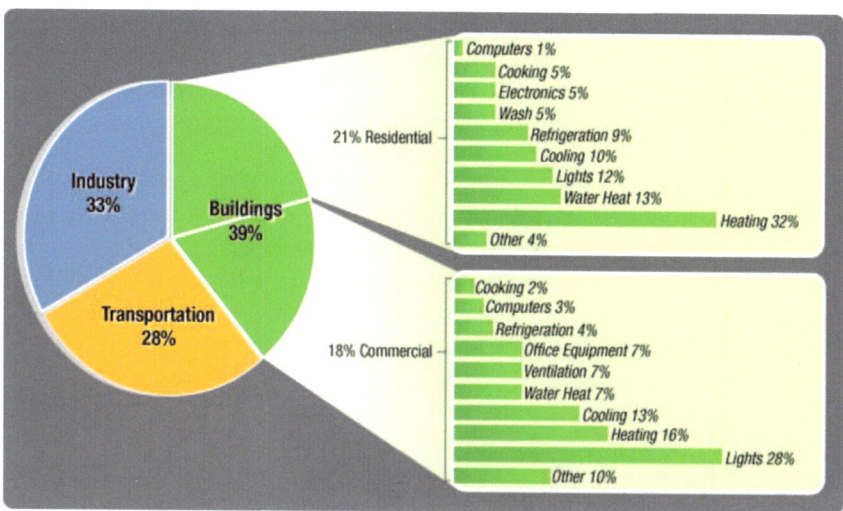

FIGURE A.4 Energy consumption in the United States, by type of use. SOURCE: Lawrence Berkeley National Laboratory.

demand. Commercial light use and residential heating make up the bulk of their respective building types, but several other smaller items make up the rest of the energy usage. Perhaps reductions in several of these "low-hanging fruit" items can contribute significantly in reducing total energy consumption.

Impediments to Changing the Energy System

Insufficient Scope and Scale of Research and Development Funding to Fuel IT-Enabled Innovation in the Electricity Sector

Challenged to consider opportunities for IT and CS research to contribute to sustainability, participants reflected on the history of IT successes and on whether those successes might offer important lessons. The enormous payoffs from IT R&D investment have been investigated by several studies of the National Research Council's Computer Science and Telecommunications Board, including *Evolving the High Performance Computing and Communications Initiative to Support the Nation's Infrastructure* (1995); *Funding a Revolution: Government Support for Computing Research* (1999); *Making IT Better: Expanding Information Technology Research to Meet*

Society's Needs (2000); and *Innovation in Information Technology* (2003).[25] These reports have shown how research partnerships between the federal government and industry ultimately led to the creation of many well-known multibillion-dollar industries. These results suggest the potential sustainability payoffs from the right investments in IT.

Many of the advancements presented in the CSTB reports, such as those in processors or networking, required significant financial investment from both industry and government. The software industry spends approximately 13.5 percent of revenues on R&D, the health care industry spends about the same, and the computer hardware industry spends about half of that.[26] By contrast, R&D spending by the electric utility sector is about 0.1 percent of revenues, perhaps due to the fact that the sector has been very stable, with little innovation or push for innovation, a context that seems to be changing rapidly.[27]

Sustainability is a large, broad-ranging problem, and apportioning limited research dollars to effective ends is a difficult challenge. One consequence of this low level of support and the resulting small number of technical researchers at utility companies is that opportunities for partnership between academic researchers and utility companies are rare.

Government funding is also limited. In 2010, the U.S. Department of Energy provided $130 million and created three different energy hubs in innovation.[28] However, a workshop attendee commented that even this amount is much smaller than would be needed if a significant shift were to be made toward sustainable energy sources or if total energy consumption were to be decreased.

Misalignment of Incentives for More Sustainable Generation and Use

The energy-utility market, as described earlier, has evolved to provide a critical resource, at low price, with supply almost instantaneously

---

[25]National Research Council, *Evolving the High Performance Computing and Communications Initiative to Support the Nation's Infrastructure*, Washington, D.C.: National Academy Press (1995); National Research Council, *Funding a Revolution: Government Support for Computing Research*, Washington, D.C.: National Academy Press (1999); National Research Council, *Making IT Better: Expanding Information Technology Research to Meet Society's Needs*, Washington, D.C.: National Academy Press (2000); National Research Council, *Innovation in Information Technology*, Washington, D.C.: The National Academies Press (2003).

[26]Jill Jusko, *R&D Spending: By the Numbers*. Industryweek.com. January 2010. Available at http://www.industryweek.com/articles/rd_spending_by_the_numbers_17988.aspx.

[27] Jusko, *R&D Spending*, 2010, available at http://www.industryweek.com/articles/rd_spending_by_the_numbers_17988.aspx.

[28]Department of Energy, "Obama Administration Launches $130 Million Building Energy Efficiency Effort," February 12, 2010, available at http://energy.gov/articles/obama-administration-launches-130-million-building-energy-efficiency-effort.

matched to demand. Although historically it required considerable innovation and tremendous capital investment to meet these constraints, there are additional market impediments to creating a more sustainable system. Perhaps the most obvious is that, generally speaking, utility companies have historically charged for usage by the kilowatt-hour, resulting in little economic incentive to reduce the number of kilowatt-hours used. regulation prevents vertical monopolies, but there is often an interest in owning an entire vertical market—one organization owning or operating both the production and the delivery systems—and extracting marginal profit mostly by locking customers in to the system. Participants observed that horizontal market stratification would help drive efficient markets. This limited-competition system means that the utility industry is not particularly motivated to shift technologies, which may drive up the cost of production in the short term. The question again is where the investment to drive new technologies is going to come from.

While the utility companies have little incentive to encourage reductions in energy use, consumers themselves have undervalued energy. As noted earlier, consumers have become accustomed to inexpensive power and also have little understanding of how power is produced and of the resulting environmental damage. Consumers have even less knowledge or easy insight into the energy costs of producing and transporting foods and goods. The energy cost, including the accompanying externalities such as environmental and social damage, is not easily re ected in the price of goods. If these costs were re ected directly in the price, more energy-efficient choices might be made.

Infrastructural and Organizational Impediments

Impediments to making progress on sustainability in addition to those discussed above include infrastructural and organizational realities. The scale of the sustainability problem is immense, and the infrastructure systems that bear on sustainability—such as energy, water, and food distribution—are just as massive. In addition, diversity of use within the system adds a level of complexity. The use and design of each building site and the water distribution and transportation system of each city have unique characteristics that make a one-size-fits-most solution impractical. Furthermore, the traditional production cycle does not apply; infrastructures are not projects that are developed, improved, and shipped; they are built once. Cities are developed over a span of 100 years or more, with refinements, changes, and "debugging" taking place little by little. Once these systems are rolled out, even if they do not function as well as they could, they become, in effect, stranded assets.

The market structure also creates impediments to better technological change. The market is highly fragmented; energy sources vary, and energy use is even more dispersed. Each industry that participates in the energy market has its own unique needs, regulatory requirements, and certification programs. Individual industries and companies create their own technology standards. Unique industry and corporate technology standards also make one-size-fits-most solutions impractical. Efforts to deploy, say, monitoring and data-collection tools in these sorts of environments are challenged. Equipment used for monitoring the use of each resource system—energy, water, food—within cities becomes difficult to build and deploy. Additionally, these monitoring devices, if built in to the initial infrastructure, need to be able to collect a wide variety of data and be sturdy enough to function over long periods of time.

Research Impediments

The critical nature of the sustainability problem and energy crisis combined with their scale and complexity often means that researchers are dedicating entire careers working to address pieces of the problem. This scale and complexity mean that choosing avenues of investigation is a high-risk proposition. If a path that a researcher follows turns out to be incorrect or a dead end, the mistake can be career ending. Furthermore, these sustainability and energy problems are inherently multidisciplinary, which adds another barrier to academic work often confined to single disciplines.

In many subfields of computer science, the ultimate goals can be defined reasonably clearly, even if the description of the goal is as simple as: Make computers faster. Well-defined goals also imply a clear definition of success. While there are some goals to work toward in addressing the sustainability problem, such as decreasing the levels of greenhouse gases in the atmosphere, they tend to be less well defined (should the focus be on lowering energy use or on the use of more sustainable energy sources?) and have less clear benchmarks for success.

Potential Computer Science Contributions

In the fourth session of the workshop, participants brainstormed about potential further contributions of computer science to sustainability. Computer science is well positioned to provide technical options that could help address some sustainability challenges. Additionally, the distinctive culture, methodologies, and approaches of computer science may shed new light on methodologies, processes, and concepts that could be useful in sustainability. Speakers discussed several such cultural attributes, including the following:

- *Culture of innovation.* Computer scientists are used to developing and deploying new tools almost constantly and to doing these things quickly. Participants argued that this exible, catch-all approach allows for broader ideas and more creativity.
- *Large-scale systems approach.* Computer scientists have experience building big things, such as massive integrated circuits, which have tens of millions of design points that need to be correct when built, and software artifacts that today measure in the millions of lines of code. Computer scientists also understand system approaches.
- *Understanding of open-information systems.* Computer scientists tend to understand the value of open systems and are often forced to engage with demands for system-level considerations such as compatibility and interoperability. Distributed grid management, ecosystems understanding, crisis and disaster response, and resource tracking and optimization can all benefit from open, interoperable information systems. With large amounts of data being collected, privacy and security become an issue, which, again, computer scientists have experience managing.[29]
- *Business transformation, often with efficiency as a goal.* As new technologies have become available, the computer science industry has transformed itself several times. For example, participants noted that data centers are drastically different now than they were just 2 years ago. This change has been driven partly by efficiency concerns. Furthermore, computer science has been fundamental in transforming other industries, for example, car ownership, media consumption, and banking, in interesting ways. Advances in smartphones, the Global Positioning System, and human-computer interaction have contributed to the success of short-term car-use services, such as Zipcar; advances in telecommunications networks and file compression have made Internet video streaming a viable alternative to the video store; and computer and information security have encouraged confidence in online banking.
- *Educating in a dynamic environment.* Because sustainability efforts are complex, multidisciplinary problems, universities will need new ways to teach scientists and engineers to resolve these problems. Computer science has historically adapted to changes in curriculum and changes in the overall technological environment by shifting teaching techniques very

---

[29]The Internet is an example in which computer science has incorporated open-information systems, shared standards, and a complex understanding of intellectual property. Although there have been questions and debates about appropriate infrastructure, standards, and intellectual property, especially as more of the Internet has been commercialized, there is still copious knowledge that can be gleaned from the computer science community on building interdisciplinary, complex systems.

rapidly. These educational tools, developed within the computer science discipline, can help develop the next wave of scientists and engineers.

Wrap-Up Discussion

This session resulted in a wide-ranging discussion from the participants at the workshop. Several key points raised are outlined below:

- Within the information technology industry, significant innovation has been accomplished at small businesses or start-ups, which then are often acquired by large corporations. This suggests that small amounts of money could fund highly innovative projects in sustainability, given the proper organizational structure and incentive.
- Although sustainability can be viewed in many ways as a technical problem, it will not be solved through technological solutions alone. Some people conjecture that in addition to major scientific and technical breakthroughs to meet sustainability challenges, large-scale social change will be needed, perhaps even on the scale of the U.S. civil rights movement. Computer scientists can contribute tools that encourage individual participation in addressing sustainability challenges.
- Small businesses often require specialized information that can be hard to acquire. Computational techniques and technologies can help by providing ways to collect, aggregate, distribute, and analyze data, as well as techniques for communication and coordination as appropriate.[30]
- There are trade-offs in discussing solutions. For example, raising temperatures in server rooms may reduce cooling loads but lead to higher failure rates. These trade-offs and failure rates have to be fully understood so that the best trade-offs can be made.
- Domain scientists (such as ecologists, transportation specialists, civil and power engineers) need to share information and knowledge with people doing innovation, including computer scientists. The first step for computer science might simply be finding a better way to present these data, which would help policy makers. Decision makers need to understand the data more clearly before they can form policy.

---

[30] An example was given of a case in which a number of local coffee shops were interested in purchasing biodegradable products. Today, biodegradable cups are more expensive and are only affordable if purchased in very large quantities; IT can link companies willing to purchase and share large shipments. Also, not all biodegradable cups are biodegradable to the same extent, an information gap that could be solved with more usable data. However, computer scientists typically have little knowledge about the chemical makeup of products, and so there is also a need for coordination across multiple disciplines and industries.

APPENDIX A

WORKSHOP AGENDA

May 26, 2010
Washington, D.C.

8:30-8:35 a.m. **Welcome**
Deborah L. Estrin, University of California, Los Angeles
Chair, Committee on Computing Research for Environmental and Societal Sustainability

8:35-10:45 a.m. **Session 1: Expanding Science and Engineering with Novel CS/IT Methods: "The Need to Turn Numbers into Knowledge"**
Committee respondent: Daniel Kammen, University of California, Berkeley

What are some example areas of efforts in sustainability and related research where the interface of disciplinary and interdisciplinary research with new methods in computer and information science can generate new innovations and knowledge? One example is the smart grid, which provides a physical and information technology medium where new levels of efficient and clean energy and information management are possible, and where new levels of data security are needed. Discussion topics range from grid management to the introduction of smart management and charging systems for low-carbon electric vehicles. Another example is ecological resilience and ecosystem function, which is the monitoring and modeling of ecological change and of the interactions related to ecological robustness and requires new tools for temporal and spatial resolution, new methods to explore the dynamics of connectivity in ecological systems, and teasing out the ranges of anthropogenic impacts.

Vijay Modi, Columbia University: "Criticality of CS and IT to Sustainability"
Robert Pfahl, International Electronics Manufacturing Initiative, Inc.: "Towards a Sustainable World Through Electronic Systems and IT"
Neo Martinez, Pacific Ecoinformatics and Computational Ecology Lab: "Numbers: Where They Come from and What to Do with Them to Live More Sustainably on Earth"
Adjo Amekudzi, Georgia Institute of Technology: "Using Social Sustainability Measures as Inputs in Planning and Design"
Thomas Harmon, University of California, Merced: "Environmental Cyberinfrastructure and Data Acquisition"

11:00 a.m.-1:00 p.m. Session 2: Understanding, Tracking, and Managing Uncertainty Throughout the Science-to-Policy Pipeline
Committee respondent: *Thomas Dietterich, Oregon State University*

Explicit representation of uncertainty is rare in the science-to-policy pipeline. Data products resulting from fusing information from multiple instruments are often treated as exact when input to models. Outputs from predictive and simulation models are often treated as exact when input to policy making. Policy optimization for management (e.g., reserve design, fishing quotas, habitat conservation plans) often is not robust to uncertainty in the problem formulation or the objectives. Uncertainty about future decision making and imperfect implementation of policies injects additional uncertainty into planning for the future.

- What are the sources of uncertainty that should be explicitly captured?
- What methods are suitable for explicitly representing uncertainty?
- Is the technological state of the art sufficient to model the many different avors of uncertainty present in large-scale sustainability problems? If not, what characterizes the types of uncertainty that are insufficiently modeled?
- What methods are suitable for assessing uncertainty in each stage of the pipeline? What shortcomings need to be addressed?
- Is the state of the art in human factors, interfaces, and CSCW (computer-supported cooperative work) sufficient to support the large-scale systems, models, and data sets that are necessary to tackle large-scale sustainability problems? If not, what needs are unmet?
- What are the appropriate techniques for working with uncertain data in data fusion, data assimilation, predictive modeling, simulation modeling, and policy optimization?
- Is a pipeline architecture sufficient, or do we need a fully coupled architecture in which policy questions can reach all the way back to guide data collection and data fusion?
- How can explicit uncertainty representations be integrated into scientific work ow tools?
- Are there alternatives to explicit uncertainty representations that can improve the robustness of management policies to all of these sources of uncertainty?

Peter Bajcsy, National Institute of Standards and Technology: "Instruments and Scientific Work ows"

Chris Forest, Pennsylvania State University: "Assessing Uncertainty in Climate Models"
David Brown, Duke University: "Robust Optimization under Uncertainty"
John Doyle, California Institute of Technology: "Theory and Methodology of Robust-yet-Fragile Systems Analysis"

1:30-3:00 p.m.   **Session 3: Creating Institutional and Personal Change with Humans in the Loop**
Committee respondent: *Alan Borning, University of Washington*

Achieving sustainability objectives demands behavioral changes at the institutional and individual levels. In designing and developing smarter systems, an important question is how to embed interfaces that work. The human-system interaction literature is replete with counterexamples and numerous failed cognitive models, serving as cautionary tales. Complicating matters, human-system interaction issues arise both with regard to individuals in homes and offices and for administrators of larger systems or facilities. Further, interactions occur at different scales—on the one hand in a day-to-day time frame for users and on the other in ways that allow incorporation of feedback from the system either to the system itself or to decision makers thinking about larger-scale resource management considerations, for example.

• How can data and information be presented at the appropriate granularity and timescale to be most effective? What system, application, and user factors bear on the human-system interaction design choices?
• Describe the potential impacts of the various contexts and trade-off decisions that might need to be made, including the impact of context (e.g., stakeholders, and so on); the impact of large versus small groups versus individuals; the impact of income; the impact of use by or for cities versus businesses versus individuals; the role of middleware, the supply chain, and so on.
• How do human factors affecting energy use drive the use and design of technology? How can this be accounted for? When are power, networking, products, and so on really needed? Discuss human choice and its impact on consumption, disposal, reuse, and so on.

Bill Tomlinson, University of California, Irvine: "Greening Through IT"
Shwetak Patel, University of Washington: "Residential Energy Measurement and Disaggregated Data"
Eli Blevis, Indiana University: "Sustainable Interaction Design"

3:15-4:00 p.m. **Session 4: Overcoming Obstacles to Scientific Discovery and Translating Science to Practice**
Committee respondent: *David Culler, University of California, Berkeley*

• What are the motivations for and impediments to applying innovative information technologies to sustainability challenges, and how do they differ by domain?
• How can large-scale science addressing real-world problems be made credible, if reproducibility is not possible?
• What lessons can be applied from the transformation of the Internet into a critical infrastructure that must avoid ossification?
• What is the appropriate mix of empiricism, innovation, and application for computer science to have an impact in the area of environmental sustainability?

David Douglas, National Ecological Observatory Network: "The Role of CS in Open, Sustainability Science"

4:00-5:00 p.m. **Capstone Session and Plenary Discussion**
Deborah L. Estrin, Committee Chair
Randal Bryant, Carnegie Mellon University

B

Biographies of Committee Members and Staff

Deborah L. Estrin (NAE), *Chair,* is a professor of computer science with a joint appointment in electrical engineering at the University of California, Los Angeles; holds the Jon Postel Chair in Computer Networks; and is a co-founder of the non-profit, Open mHealth. Professor Estrin received her Ph.D. (1985) in computer science from the Massachusetts Institute of Technology and her B.S. (1980) from the University of California, Berkeley. Her early research (conducted while she was on the Computer Science Department faculty at the University of Southern California [USC] and the USC Information Sciences Institute) focused on the design of network and routing protocols for very large, global networks, including multicast routing protocols, self-configuring protocol mechanisms for scalability and robustness, and tools and methods for designing and studying large-scale networks. From 2002 to 2012 she founded and directed the multidisciplinary, National Science Foundation (NSF)-funded Science and Technology Center for Embedded Networked Sensing (CENS), which developed environmental monitoring technologies and applications (http://cens.ucla.edu). Currently Professor Estrin explores participatory sensing and mHealth systems, leveraging the programmability, proximity, and pervasiveness of mobile devices; deployment contexts include health (http://openmhealth.org), community data gathering, and education (http://mobilizingcs.org). Professor Estrin has been a co-principal investigator on numerous NSF and Defense Advanced Research Projects Agency (DARPA)-funded projects and has been an active participant in several government-sponsored studies. She chaired a 1997-1998 DARPA Information Science and Technology Study Group study on sensor networks, and the 2001 National Research Council (NRC) study on networked embedded computing, which produced the report *Embedded, Everywhere: A Research Agenda for Networked Systems of Embedded Computers*. She later chaired the Sensors and Sensor Networks subcommittee of the NEON (National Ecological Observatory Network) Design Committee (www.neoninc.org). Professor Estrin also served on the Advisory Committees for the NSF Computer and Information Science and Engineering (CISE) and Environmental Research and Education (ERE) Directorates, and is a former member of the NRC's Computer Science and Telecommunications Board. She was an editor of the *IEEE/ACM Transactions on Networking*, and a program committee member for many networking-related conferences, including Sigcomm (Special Interest Group on Data Communication), Infocom (International Conference on Computer Communications), MobiCom, and MobiSys. She was the steering group chair and general co-chair for the first Association for Computing Machinery (ACM) Conference on Embedded Networked Sensor Systems, Sensys 2003, and served as one of the first associate editors for the *ACM Transactions on Sensor Networks*. Professor Estrin is a fellow of the ACM, the American Association for the Advancement of Science, and the Institute of Electrical and Electronics Engineers. She was selected as the first ACM-W Athena Lecturer in 2006, was awarded the Anita Borg Institute's Women of Vision Award for Innovation in 2007, was inducted into the Women in Technology International Hall of Fame in 2008, and awarded Doctor Honoris Causa from École Polytechnique Fédérale de Lausanne in 2008. Professor Estrin was elected to the American Academy of Arts and Sciences in 2007 and into the National Academy of Engineering in 2009.

Alan Borning is a professor in the Department of Computer Science and Engineering at the University of Washington, an adjunct faculty member in the Information School, and a fellow of the Association for Computing Machinery. He received a B.A. in mathematics from Reed College (1971) and an M.S. (1974) and a Ph.D. (1979) from Stanford University. His principal research interests are in human-computer interaction and designing for human values. His current research projects include online tools to support civic engagement and participation, mobile tools to aid transit riders, and designing systems to support more effective public participation in land use and transportation deliberations, supported by sophisticated simulation data. Earlier he worked on programming languages and UI (user interface) toolkits, including constraint-based languages and systems and on object-oriented languages.

David Culler (NAE), a professor and chair of computer science, associate chair of electrical engineering and computer sciences, and faculty director of i4energy at the University of California, Berkeley, received his B.A. from the University of California, Berkeley (1980) and an M.S. and a Ph.D. from the Massachusetts Institute of Technology (1985 and 1989, respectively). He joined the Department of Electrical Engineering and Computer Science faculty in 1989, where he holds the Howard Friesen Chair. He is a member of the National Academy of Engineering, a fellow of the Association for Computing Machinery (ACM), and an Institute of Electrical and Electronics Engineers fellow; he was selected for ACM's Sigmod Outstanding Achievement Award, and was named in the *Scientific American* Top 50 Researchers and the Technology Review: 10 Technologies That Will Change the World. He was awarded the National Science Foundation (NSF) Presidential Young Investigator Award in 1990 and the NSF Presidential Faculty Fellowship in 1992. He was the principal investigator (PI) of the Defense Advanced Research Projects Agency network embedded systems technology project that created the open platform for wireless sensor networks based on TinyOS, a co-founder and the chief technology officer of Arch Rock Corporation, and the founding director of Intel Research, Berkeley. He has done seminal work on networks of small, embedded wireless devices, planetary-scale internet services, parallel computer architecture, parallel programming languages, and high-performance communication, including TinyOS, PlanetLab, Networks of Workstations (NOW), and Active Messages. He has served on technical advisory boards for several companies, including People Power, Inktomi, ExpertCity (now Citrix Online), and DoCoMo USA. He is currently focused on utilizing information technology to address the energy problem and is co-PI on the NSF Cyber-Physical Systems projects LoCal and ActionWebs.

Thomas Dietterich, professor at Oregon State University (OSU), focuses on interdisciplinary research at the boundary of computer science, ecology, and sustainability policy. He is the principal investigator (with Carla Gomes of Cornell University) of a 5-year National Science Foundation (NSF) Expedition in Computational Sustainability. He is part of the leadership team for OSU's Ecosystem Informatics programs, including an NSF Summer Institute in Ecoinformatics. Dr. Dietterich received his A.B from Oberlin College (1977), M.S. from the University of Illinois (1979), and Ph.D. from Stanford University (1984). He is professor and director of Intelligent Systems in the School of Electrical Engineering and Computer Science at OSU, having joined the faculty there in 1985. In 1987, he was named a Presidential Young Investigator for the NSF. In 1990, he published, with Dr. Jude Shavlik, the book entitled *Readings in Machine*

Learning, and he also served as the technical program co-chair of the National Conference on Artificial Intelligence (AAAI-90). From 1992 to 1998 he held the position of executive editor of the journal *Machine Learning*. He is a fellow of the Association for the Advancement of Artificial Intelligence (1994), the Association for Computing Machinery (2003), and the American Association for the Advancement of Science (2007). In 2000, he co-founded the free *Journal of Machine Learning Research* and he is currently a member of its editorial board. He served as technical program chair of the Neural Information Processing Systems (NIPS) conference in 2000 and as general chair in 2001. He is past president of the International Machine Learning Society (IMLS) and a member of the IMLS board, and he also serves on the advisory board of the NIPS Foundation.

Daniel Kammen is the Class of 1935 Distinguished Professor of Energy at the University of California, Berkeley, with parallel appointments in the Energy and Resources Group, the Goldman School of Public Policy, and the Department of Nuclear Engineering. He serves as an Environment and Climate Partnership for the Americas (ECPA) Fellow for Secretary of State Hillary Clinton. Dr. Kammen is the founding director of the Renewable and Appropriate Energy Laboratory (RAEL), co-director of the Berkeley Institute of the Environment, and director of the Transportation Sustainability Research Center. He has founded or is on the board of more than 10 companies and has served the State of California and the U.S. government in expert and advisory capacities. Dr. Kammen was educated in physics at Cornell University and Harvard University, and held postdoctoral positions at the California Institute of Technology and Harvard. He was assistant professor and chair of the Science, Technology and Environmental Policy Program at the Woodrow Wilson School at Princeton University before moving to the University of California, Berkeley. Dr. Kammen has served as a contributing or coordinating lead author on various reports of the Intergovernmental Panel on Climate Change (IPCC) since 1999. The IPCC shared the 2007 Nobel Peace Prize. During 2010-2011, Dr. Kammen served as the World Bank Group's chief technical specialist for renewable energy and energy efficiency. In this newly created position to which he was appointed in October 2010, he provided strategic leadership on policy, technical, and operational fronts. The aim is to enhance the operational impact of the World Bank's renewable energy and energy-efficiency activities while expanding the institution's role as an enabler of global dialogue on moving energy development to a cleaner and more sustainable pathway. He has authored or co-authored 12 books, written more than 250 peer-reviewed journal publications, testified more than 40 times at U.S. state and federal congressional briefings, and has provided various governments with more than 50 technical reports. Dr. Kammen also served for many years on the Technical Review Board of the Global Environment Facility. He is a frequent contributor to or commentator in international news media, including *Newsweek, Time, The New York Times, The Guardian,* and *The Financial Times*. Dr. Kammen has appeared on *60 Minutes* (twice), *Nova,* and *Frontline,* and he hosted the six-part Discovery Channel series *Ecopolis*. He is a permanent fellow of the African Academy of Sciences and a fellow of the American Physical Society. In the United States he serves on a board and a panel of the National Academy of Sciences.

Jennifer Mankoff is an associate professor in the Human-Computer Interaction Institute at Carnegie Mellon University. She earned her B.A. at Oberlin College and her Ph.D. in computer science at the Georgia Institute of Technology. Her research embodies a human-centered perspective on data-driven applications. Her goal is to combine empirical methods with technological innovation to construct middleware (tools and processes) that can enable the creation of impactful data-driven applications. Examples of such application areas include sensing and in uencing energy-saving behavior, web interfaces for individuals with chronic illness, and assistive technologies for people with disabilities. Dr. Mankoff helped found the sustainable-chi group (www.sustainable-chi@ googlegroups.com). Her research has been supported by Google, Inc., the Intel Corporation, IBM, Hewlett-Packard, Microsoft Corporation, and the National Science Foundation. She was awarded the Sloan Fellowship and the IBM Faculty Fellowship.

Roger D. Peng is an associate professor of biostatistics at the Johns Hopkins Bloomberg School of Public Health. He received his Ph.D. in statistics from the University of California, Los Angeles. He is a prominent researcher in the areas of air pollution and health risk assessment and statistical methods for spatial and temporal data. Dr. Peng is a national leader in the area of methods and standards for reproducible research; he is the Reproducible Research editor for the journal *Biostatistics*. He has developed novel approaches to integrating complex national databases for assessing population health effects of environmental exposures and has developed software for efficiently distributing data over the web for disseminating reproducible research. Dr. Peng's research is highly interdisciplinary; his work has been published in major substantive and statistical journals, including the *Journal of the American Medical Association, Journal of the American Statistical Association, Journal of the Royal Statistical Society,* and *American Journal of Epidemiology*. Dr. Peng is the author of more than a dozen software packages implementing statistical methods for environmental studies, methods for reproducible research, and data distribution tools. He has also given workshops, tutorials, and short courses in statistical computing and data analysis.

Andreas Vogel is vice president in the global business incubator at SAP Labs in Palo Alto, California, where he currently works on virtual economies and recommender systems for online games. Previously he incubated a product for analyzing smart meter data and developed the next generation of sustainability-related software solutions, including those involving carbon accounting, energy management, and electrified vehicles. He also helped to create and implement SAP's sustainability strategy. Before joining SAP, Dr. Andreas held various research, technology, and business positions around the world—among them, chief scientist at Borland and chief technology officer and co-founder of Mspect, where he developed monitoring solutions for mobile data networks. Dr. Andreas holds an M.Sc. and a Ph.D. in computer science from Humboldt University, Berlin, Germany. He co-authored four books on Common Object Request Broker Architecture (CORBA), Enterprise Java Beans, and enterprise resource planning (ERP) published by J. Wiley and Sons.

Staff

Lynette I. Millett is a senior program officer and study director at the Computer Science and Telecommunications Board (CSTB), National Research Council of the National Academies. She currently directs several CSTB projects, including an exploration of foundational science in cybersecurity. She recently completed the project that produced *Strategies and Priorities for Information Technology at the Centers for Medicare and Medicaid Services*. Ms. Millett's portfolio includes substantial portions of CSTB's recent work on software, identity systems, and privacy. She directed, among other projects, those that produced *The Future of Computing Performance: Game Over or Next Level?*, an examination of the causes and implications of the slowdown in the historically dramatic exponential growth in computing performance; *Software for Dependable Systems: Sufficient Evidence?*, an exploration of fundamental approaches to developing dependable mission-critical systems; *Biometric Recognition: Challenges and Opportunities*, a comprehensive assessment of biometric technology; *Who Goes There? Authentication Through the Lens of Privacy*, a discussion of authentication technologies and their privacy implications; and *IDs—Not That Easy: Questions About Nationwide Identity Systems*, a post-9/11 analysis of the challenges presented by large-scale identity systems. She has an M.Sc. in computer science from Cornell University, where her work was supported by graduate fellowships from the National Science Foundation and the Intel Corporation; and a B.A. with honors in mathematics and computer science from Colby College.

Virginia Bacon Talati is an associate program officer at the Computer Science and Telecommunications Board, National Research Council of the National Academies. She formerly served as a program associate with the Frontiers of Engineering program at the National Academy of Engineering. Prior to her work at the Academies, she served as a senior project assistant in education technology at the National School Boards Association. She has a B.S. in science, technology, and culture from the Georgia Institute of Technology and an M.P.P. from George Mason University with a focus in science and technology policy.

Shenae Bradley is a senior program assistant at the Computer Science and Telecommunications Board, National Research Council of the National Academies. She has provided support for the Committee on Sustaining Growth in Computing Performance, the Committee on Wireless Technology Prospects and Policy Options, and Computational Thinking for Everyone: A Workshop Series Planning Committee, to name a few. Previously, she served as an administrative assistant for the Ironworker Management Progressive Action Cooperative Trust and managed a number of apartment rental communities for Edgewood Management Corporation in the Maryland/DC/Delaware metropolitan areas. Ms. Bradley is in the process of earning her B.S. in family studies from the University of Maryland at College Park.